THE MAGIC OF NUMBERS

ERIC TEMPLE BELL

DOVER PUBLICATIONS, INC.
New York

Published in Canada by General Publishing Company, Ltd., 30 Lesmill Road, Don Mills, Toronto, Ontario.
Published in the United Kingdom by Constable and Company, Ltd., 3 The Lanchesters, 162–164 Fulham Palace Road, London W6 9ER.

This Dover edition, first published in 1991, is an unabridged and unaltered republication of the work first published by Whittlesey House/McGraw-Hill Book Company, Inc., New York, in 1946.

Manufactured in the United States of America
Dover Publications, Inc., 31 East 2nd Street, Mineola, N.Y. 11501

Library of Congress Cataloging-in-Publication Data

Bell, Eric Temple, 1883–1960.
The magic of numbers / Eric Temple Bell.
p. cm.
Reprint. Originally published: New York : Whittlesey House. McGraw-Hill, 1946.
ISBN 0-486-26788-1 (pbk.)
1. Number concept. 2. Mathematics—Philosophy. 3. Mathematics—History. I. Title.
QA141.B39 1991
512′.7—dc20 91-18322
CIP

Note

The topic of this book is described in the first few pages of the opening chapter. It concerns what may be the least expected turn of scientific thought in twenty-five centuries. Should this return to a remote past—for that is what the most recent philosophy of science really is—be generally accepted, our descendants a few generations hence will look back on us and our science as incredibly unenlightened.

Not much of the proposed substitute for the scientific method as commonly understood has been discussed outside professional scientific circles. An untechnical account of the origins and progress of the new approach to nature may therefore be of interest to those who do not make their livings at science. It will appear that the new and the old are strangely alike.

For valuable criticisms and suggestions I am indebted to many friends, professional and other. Though I alone am responsible for what finally got written down, I should like especially to thank Eleanor Bohnenblust, Fréderic Bohnenblust, Mary Mayo, and Lasló Zechmeister for their patience and helpfulness with it all, and Nina Jo Reeves for preparing the manuscript for publication. For permission to reprint the excerpts that appeared in *Scripta Mathematica,* I am indebted to the editors of *Scripta.*

E. T. Bell

Contents

CHAPTER	PAGE
1. The Past Returns	1
2. A Royal Mace	13
3. For Their Own Sake	24
4. The Decisive Century	32
5. A Difference of Opinion	38
6. Wisdom as a Profession	44
7. Not Much, But Enough	56
8. One or Many?	68
9. A Dream and a Doubt	74
10. Half Man, Half Myth	80
11. Discord and Harmony	97
12. Harmony and Discord	115
13. Mythology Transformed	135
14. The Cosmos as Number	148
15. Himself Made It?	194
16. Intimations of the Infinite	201
17. A Miscarriage of Reason	220
18. Politics and Geometry	228
19. "Another I"	246
20. Number Deified	258
21. Pythagoras in Purgatory	276

Contents

CHAPTER PAGE

22. Saints and Heretics 312
23. A Turning Point 330
24. The Skeptical Bishop 336
25. Believer and Disbeliever 345
26. Changing Views 362
27. Return of the Master 387

CHAPTER - - - - - - - - - - - - - - 1

The Past Returns

THE hero of our story is Pythagoras. Born to immortality five hundred years before the Christian era began, this titanic spirit overshadows western civilization. In some respects he is more vividly alive today than he was in his mortal prime twenty-five centuries ago, when he deflected the momentum of prescientific history toward our own unimagined scientific and technological culture.

Mystic, philosopher, experimental physicist, and mathematician of the first rank, Pythagoras dominated the thought of his age and foreshadowed the scientific mysticisms of our own. So varied was his genius that the crassest superstitions and the most uncompromising rationalisms might appeal to his authority—"Himself said it"—all down the Middle Ages. The essence of his teaching was the mystic doctrine that "Everything is number." With Galileo's revival in the late sixteenth century of the experimental method in the physical sciences, a method in which Pythagoras had pioneered nearly twenty-two centuries earlier, number mysticism passed out of science.

The seventeenth century saw the creation by Newton and Leibniz of a new mathematics, devised to bring the continuously varying flux of nature under the domination of rigorous reasoning. Combining this dynamic mathematics with precise observation and purposeful experiment, Newton and his followers in the seventeenth and eighteenth centuries fixed the modern scientific method for the astronomical and physical sciences. The form they gave it was to stand unchallenged till the third decade of the twentieth century.

The aim of this method was twofold: to sum up the observable phenomena of the physical universe in readily apprehended generalizations—called by their inventors or discoverers "laws of nature"; to enable human beings in some degree to predict the course of nature. Always observation and experiment were the first and last court of appeal. However reasonable or however inevitable the verdict of mathematics or other strict deductive reasoning might appear, it alone was not accepted without confirmation by this final authority.

The successes of the method heavily overbalanced its failures all through the nineteenth century and well into the twentieth. In less than two centuries the application of scientific technology to industry wrought a profounder transformation of western civilization than had all the wars and social upheavals of the preceding four thousand years.

Concurrently with this vast revolution in the material world, equally subversive changes from time to time overthrew established creeds that had possessed the thoughts of men for scores of generations. The universe disclosed by science was not always that of revelation and tradition, nor even that which a supposedly infallible logic insisted must be the fact. Here also the absolutes of more than two thousand years were impartially scrutinized. Those that had proved barren of positive knowledge were ruthlessly abandoned. The unaided reason as an implement of discovery and understanding in the exploration of the material universe dropped out of use. Its sterility in science then cast suspicion on it in its own traditional domain. Of what human value were truths immune to any objective test that human beings might invent? Protests that truths other than those of science exist timelessly in a realm of Eternal Being, and are forever inaccessible to the finite reach of science, were silenced by the dictum, "Experiment answers all." Then, quite suddenly, about the year 1920, the most positive of all the sciences began to hesitate.

By the middle 1930's a few prominent and respected physicists and astrophysicists had reversed their position squarely. Facing the past unafraid, they strode boldly back to the sixth century B.C. to join their master. Though the words with which they greeted him were more sophisticated than any that Pythagoras might have uttered, they were still in his ancient tongue. The meaning implicit in their refined symbolisms and intricate metaphors had not changed in twenty-five centuries: "Everything is number." He understood what they were saying.

The retreat from experiment to reason was applauded by some philosophers and scientists, deplored by others. But the fact in the new movement was beyond dispute. Either the leaders had gone back to Pythagoras to acknowledge that he had been right all these centuries, or he had come forward to convince them that the modern scientific method of Galileo and Newton is a delusion.

On a first, exploratory pilgrimage to the past the daring ultramoderns had lingered for a few moments in the shadow of Plato. Quickly realizing that in all matters pertaining to the mysteries of number here was only the pupil, they sought his master. Two centuries before Plato was born, Pythagoras had believed and taught that the pure reason alone can reveal the truth of anything; observation and experiment are snares to trap and betray the unwary senses. And of all languages in which constant knowledge as opposed to variable opinion may be described, that of number is the only one on which the pure reason may safely rely. "Himself said it," and now, twenty-five centuries after his historic death, he was repeating himself in the language of a nascent science.

A devout believer in the doctrines of reincarnation and the transmigration of souls, Pythagoras may at last have found a congenial habitation in the sheer abstractions of twentieth century theoretical physics. "For my own lapses from the one true faith," he may now reflect, "I was condemned to spend life after life in the vile dogmas of false philosophers and in the viler imaginings of base numerologists. But now I am unbound from the Wheel of Birth. When I experimented with my hands and my hearing to discover the law of musical intervals, I sinned against the eternal spirit of truth, defiling my soul with the unclean things of the senses. Then I beheld the vision of Number, and knew that I had betrayed my better part. By proclaiming the truth that everything is number, I sought to cleanse my soul and gain release from the Wheel. But it was not enough. Few believed and many misunderstood. To expiate my transgression I passed through that purgatory of error and falsehood, a name honored in the mouths of fools. Now I discern an end to my torment in the dawn of a new enlightenment which was already old ages before I was Pythagoras. The deceptions of the senses shall mislead mankind no more. Observation and experiment, the deceitful panders of sensory experience, shall pass from human memory and only the pure reason remain. Everything is number."

The master's prophecy becomes less abstract and closer to the scientific actualities of the twentieth century. Speaking as a mathematical physicist and mathematical astrophysicist he proceeds to details. "I believe . . . that all the laws of nature that are usually classed as fundamental can be foreseen wholly from epistemological considerations." In a brief aside he reminds us that epistemology is that department of metaphysics which deals with the theory of human knowledge. To preclude any possible misunderstanding of his meaning he elaborates his heretical creed. "An intelligence unacquainted with our universe, but acquainted with the system of thought by which the human mind interprets to itself the content of its sensory experience, should be able to attain all the knowledge of physics that we have attained by experiment. He would not deduce the particular events and objects of our experience, but he would deduce the generalizations we have based on them. For example, he would infer the existence and properties of sodium, but not the dimensions of the earth."

If Pythagoras—ventriloquizing thus in 1935 through Sir Arthur Eddington, a leader in the retreat to the past—should be right, it would seem that the experimental scientists since Galileo and Newton have gone to much unnecessary labor to discover the obvious and proclaim it in truisms. If it is false that experiment answers all, it may be true, as some of the ancients believed, that reason answers all, or, as the successors of Pythagoras seem to believe, nearly all. For, as we have just been cautioned, reason may be unable to deduce the diameter of the earth from any data wholly within the human mind. But this defect is entirely negligible in comparison with the ability to foresee the existence and properties of the chemical elements "wholly from epistemological considerations."

By taking sufficient thought the scientific epistemologist may rediscover for himself, without once rising from his chair in an otherwise empty room, all that three centuries of observation and experiment since Galileo and Newton have taught us of the "fundamental laws" of mechanics, heat, light, sound, electricity and magnetism, electronics, the constitution of matter, chemical reactions, the motions of the heavenly bodies, and the distribution in space of stellar systems. And by the same purely abstract considerations the thoughtful epistemologist may attain verifiable knewledge of natural phenomena which are still obscure to science, for example, the internal motions of the spiral nebulae.

Should only some of these impressive claims be sustained, the twentieth century return to Pythagoreanism may be remembered ten thousand years hence as the dawn of a lasting enlightenment and the end of the long night of error which descended on western civilization in the seventeenth century. The costly apparatus of our laboratories and observatories will have crumbled and rusted away, except possibly for a few relics fearfully preserved in the World Museum of Human Error. Above the entrance the guardians of public sanity will have inscribed the truths that liberated mankind: "Experiment answers nothing. Reason answers all." To balance these, the same guardians will have embellished the pediment of the Temple of Knowledge and Wisdom with the summary of the universe and a solemn warning: "Everything is Number. Let no man ignorant of Arithmetic enter my doors."

But all this is in the calm certainty of a golden age of the future while we, unhappily, must endure the steel and errors of the present. To mitigate our lot we may return to the past for an hour or two, to read in it the certainty of our present and the hope of our future.

What shall we ask the past? Numerous interesting questions suggest themselves. How did human beings like ourselves ever come to believe the nonsense they did about numbers? And what can have induced reputable scientists of the twentieth century A.D. to fetch their philosophy of science from the sixth century B.C.? Have the numerologists —the number mystics—been right all these centuries and the majority of thinking men wrong?

As to the origin of it all, it began some twenty-six centuries ago in the simplest arithmetic and the most elementary school geometry. None of this is beyond the understanding of a normal child of twelve. As for who may be right and who wrong, a physicist or an engineer usually is more easily seduced than a mathematician or a logician by a mathematical demonstration. Few engineers or physicists would devote their best thought to a small but incisive treatise on the unreliability of the principles of logic. It took a mathematician to do that. Logic in its most reliable form is called pure mathematics; and though mathematical reasoning, like any other, has its drastic limitations, it is still the most powerful known. But because mathematics seems to create something out of nothing, whereas it does not, superhuman powers have been ascribed to it, even by logicians and mathematicians.

When a complicated mathematical argument ends in a spectacular prediction, subsequently verified by observation or experiment, a physicist may be excused for feeling that he has participated in a miracle. And when a skilled mathematician astounds himself with a discovery he had no conscious intention of striving after, he may well believe for a few moments as Pythagoras believed all his life, and may even repeat—after the eminent English mathematician, G. H. Hardy—the following confession of faith. "I believe that mathematical reality lies outside us, that our function is to discover or *observe* it, and that the theorems which we prove, and which we describe grandiloquently as our 'creations,' are simply our notes of our observations. This view has been held, in one form or another, by many philosophers of high reputation from Plato onwards. . . ."

On coming out of the daze at his own brilliance the average twentieth-century pure mathematician might begin to doubt at least the practicality of this Platonic creed, especially if he happened to be aware of what has taken place in the philosophy of mathematics since the close of the nineteenth century. The doubter might even agree with the distinguished American geometer, Edward Kasner, that the "Platonic reality" of mathematics was abandoned long ago by unmystical mathematicians, and marvel with him that rational human beings could ever have believed anything of the kind. As he puts it, "We have overcome the notion that mathematical truths have an existence independent and apart from our minds. It is even strange to us that such a notion could ever have existed. Yet this is what Pythagoras would have thought—and Descartes, along with hundreds of other great mathematicians before the nineteenth century. Today mathematics is unbound; it has cast off its chains. Whatever its essence, we recognize it to be as free as the mind, as prehensile as the imagination."

It is not for us to judge between the two schools of thought. We note only that each of the authorities cited as a witness to the truth of mathematics published his conclusions in 1940. Even in a court of law it would be difficult to find a sharper disagreement between competent experts. A like irreconcilable difference of opinion severs the modern Pythagorean scientists from those of the older school, who still believe that reliable knowledge of the physical universe cannot be attained without observation and experiment.

My sole objective in the following chapters will be to see how these differences of opinion came about. Though the theme is number, no mathematics beyond the simplest arithmetic is required for following the story. An occasional allusion to some obvious statement about straight lines, such as young children are taught in school, need not terrify anyone if it is called geometry. The important things are not these trivialities of a grade-school education. What matters is the weird nonsense people no less intelligent than ourselves inferred from these trivialities. To prevent our excursion into the past from degenerating into a journey through a valley of dry bones, we shall become as well acquainted as we may with the great men primarily responsible for our present widely divergent opinions. The majority of the men cited are famous and their major contributions to civilization well known. The aspect of their work which is of interest here may be less familiar, though it was no less important for them than the things for which they are commonly remembered. A few names may be new to some. They are only about ten out of hundreds who left their mark on number mysticism and all that it implies for our own attempt to think straight.

Those who have had no occasion to examine for themselves what the ancient lore of numbers has done, and is doing, for their thinking habits, may be interested to linger for a while at the principal shrines where the magic of numbers paused on its way from the past to the present. Time and the continual changes in the meanings of words have confused the historical record till the hard core of arithmetical fact at the center of some ancient wisdoms is not always evident at a casual glance. Much of the influence of such apparently trivial statements as "three and seven make ten" on philosophic, religious, and scientific thinking is crusted over with the symbolisms of outmoded attempts to fabricate a meaningful image of the material universe. Ambitious and inspiring as such efforts may have been, they are far surpassed—at least in ambition—by some of the earlier struggle to explain human values in terms of numbers. Virtue to the highly imaginative Pythagoreans of antiquity was one number, vice another; and the elusive concepts of the True, the Beautiful, the Good were sublimated into "Ideal Numbers" by no lesser a metaphysician than Plato. And if it seems strange that Pythagoras should have believed that love and marriage are governed by numbers, we have but to observe the like today.

Step by step the immemorial magic of numbers has kept pace with unmystical science all down the centuries. If the patient investigation of numbers has aided the development of science and furthered such enlightenment as science can give, it has also perpetuated older beliefs that but few tolerant men would call enlightened. Once of scientific certitude, these stubborn superstitions long ago ceased to have any meaning for the literate. But the belief that number is the ultimate answer to all the riddles of the physical universe, though subtly disguised, is still recognizable in the refined mathematical mysticism of the modern Pythagoreans. Our principal concern will be to retrace the main steps by which this overwhelming conclusion has reached the living present from a past so remote that only rumors of its existence survive.

To anticipate slightly, three types of mind have been lured into comprehensive theories of life and the universe by the deceptive harmonies of numbers. Contrary to what common sense might predict, mathematicians were not the first but the last to take numbers seriously, perhaps too seriously. Behind every mathematician in the dawn of numerical thinking was a scientist, and behind every scientist a priest. The scientist may have been only a primitive astrologer who read into the wanderings of the planets more than any astronomer has yet discerned. Still, he was a scientist in that he attempted to reduce his crude observations of nature to a rational system.

To the priest looking over the scientist's shoulder the irrepressibly prolific numbers repeated a familiar tale. He and his kind had known for centuries that the most potent of all magics resides in numbers. But it was not until the common run of mankind had accepted number as an almost universal convenience in astrology, in trade, in agriculture, in astronomy, and in primitive engineering, that men who today would be recognized as mathematicians arrived and began to study numbers for their own sake. Their contribution to the stock of reliable knowledge provided more imaginative men with an inexhaustible store of curious relations between numbers to interpret as they would. The outcome was a golden age of Greek philosophy.

By the time some of the more versatile interpreters had elaborated their theories of truth and the material universe, the plebeian ancestry of the noblest doctrines of certain aristocratic philosophies had been forgotten. What was reputable arithmetic then became the exclusive possession of mathematicians and scientists. Simultaneously the old magic of numbers passed into the hands of sincere but deluded zealots, whose intentions no doubt were good, but whose sacerdotal juggling with the trivialities of arithmetic was barely distinguishable from conscious charlatanry.

With the advance of experimental science in the seventeenth century the ancient magic of numbers gradually became disreputable. It then sank almost completely out of sight in philosophy, though Kant in the late eighteenth century had some of it, and the very positive Comte about fifty years later almost lost his philosophic reason in the vagaries of numerology. What remained of it throve rankly in such dubious occupations as fortune telling. But never was its less fantastic part quite dead. Then suddenly in the third decade of the twentieth century the period of suspended animation ended. Resplendent and respectable in the dazzling symbolism of a brilliant new physics, the ancient magic of numbers rose again to vigorous life. Number returned as the ruler of an infinitely vaster cosmos than all the cramped heavens Pythagoras and Plato ever imagined. Executing an abrupt about-face, the modern Pythagoreans marched back to salute their master and offer him the augmented tribute of his own.

CHAPTER - - - - - - - - - - - - - - 2

A Royal Mace

UNTIL adequate food, clothing, and shelter are reasonably secure only the hardiest souls have leisure to ponder over man's place in the universe. It is not surprising therefore to find the utilitarian motive predominant in by far the greater part of the earliest work in numbers of which there is definite record. The Egyptian farmer of five or six thousand years ago, for example, needed to know when the annual inundation of the Nile valley could be expected, and for this a fairly reliable calendar was a necessity.

Even the crudest calendar presupposes a familiarity with numbers far beyond that attained by all but the most advanced of primitive peoples. The art of counting was not perfected in a day, and many a semi-civilized tribe has stopped short of ten in its efforts to enumerate its possessions. For such peoples all numbers greater than half a dozen or so are indistinguishable from one another and blur in an unexplored vastness. They are of no more practical importance to a homeless nomad than infinity is to a Wall Street accountant.

Instead of the modern mathematician's "infinity," the wise man of a small tribe groping toward counting contents himself with an equally nebulous "many." This is sufficiently accurate for his magical predictions: the margin between starvation and plenty is as adequately covered by the difference between six and ten as it is by the vaster unknown between ten and fifteen. By eye rather than by intellect the seer who is just a shade more observant than the herdsman senses whether the tribe has enough; it is immaterial whether it has too much.

It is unlikely that we shall ever know when, where, or how human beings first learned to count with the unthinking facility of a civilized child of seven. Nor is it probable that we shall discover what people first mastered the art of counting in all its freedom.

Admitting only tangible evidence, we can assert positively that by 3500 B.C. the Egyptians had far outgrown the primitive inability to think boldly in terms of large numbers. A royal mace of about that time records the capture of 120,000 human prisoners, 400,000 oxen, and 1,422,000 goats. These very impressive round numbers suggest one of two things. Either the victorious monarch had an active imagination and an inflated ego, or the Egyptian tally keepers had learned to estimate large collections by multiplying the number of individuals in an accurately counted sample by the guessed total number of such samples.

But even this remarkable feat and others almost as spectacular do not indicate that the Egyptians of 3500 B.C. were aware that the sequence of numbers 1, 2, 3, 4, 5 . . . is indeed endless. They may have believed subconsciously that it is always possible to conceive a number greater by one than any imagined number, but they did not put their belief on record. For anything we know to the contrary the Egyptians may have believed that the numbers 1, 2, 3 . . . come somewhere, sometime, to an end. A cast of thought more subtle than theirs had to evolve before the concept of an infinite collection could become a commonplace of mathematics and philosophy.

The 120,000 prisoners, 400,000 oxen, and 1,422,000 goats of the royal mace do, however, reveal a fact of cardinal importance in the evolution of numbers. We who learn to count glibly before we can read may have overlooked the only thing of deep significance about numbers in the entire process. This must have taken almost superhuman penetration to see when it was first observed; and it is a fair guess that very few of even the most alert observers would notice it in the conqueror's boastful cataloging of his loot. As with some other fundamentals of mathematics and science, the difficulty with this one is its apparently trivial simplicity—once it is pointed out.

Looking over his human captives and the rest, what could the victor say about each of the three groups that would be true for all? He might have observed that all three were composed of living individuals. Probably he did; but if so he did not consider the observation to be of sufficient importance to merit preservation on a ceremonial mace. Actually what he noticed and recorded was that all three of the groups—human beings, oxen, goats—could be compared by one and the same process. They could all be counted.

If this seems too trivial, we may try to imagine some characteristic other than the number assigned to each of the groups that would be equally significant and as potentially useful. The required characteristic is to be wholly independent of the natures of the individuals composing the several groups. Perhaps this is too easy; to state the problem in all its generality, what is it that all of several collections of any material things whatever have in common? Each collection is countable. Moreover, as the conqueror doubtless knew, it makes no difference to the final tally in what order the things are counted, or whether the counting is done by ones, or by sevens and ones, or by tens and ones; the outcome will always be the same. The conqueror's magicians might convince their lord that one mace could become two. But they could not have shown him 1,422,001 goats by counting only 1,422,000.

The deceptive simplicity of counting conceals the very things that have made it useful and philosophically suggestive. To give them names, these may be called the universality and the invariance of the numbers generated by counting. Universality—the always true, the always relevant—has been a goal of many philosophies. Invariance—changelessness in the midst of change—has been the quest of more than one religion, and in our own century has helped to codify the laws of the physical sciences. To take an example from everyday experience, any five persons, say, meet and part. Whatever they may do, however widely they may scatter over the earth, and however diverse their fortunes, the "five" that numbered them remains unchanged. It is independent, as nothing else in their lives may be, of the accidents of space and time. Moreover the same "five" would enumerate the individuals in any group of any five things whatever.

Commonplace to us, the universality and invariance of numbers were many centuries beyond the imagination of the stewards who counted the captives. Numbers were useful to them, and that was all they needed to know in order to survive and prosper. The origins of counting were so remote, and their own civilization so far advanced, that it probably never occurred to them to ask what a number is, or to speculate on how human beings ever chanced to invent numbers. All such troublings of the spirit were thousands of years in the future. Not even the inquisitive Greeks asked explicitly what numbers are, though Pythagoras and his followers occasionally spoke of them as if they were alive.

The other question, as to who invented numbers, may be improperly posed. It is conceivable that numbers were never deliberately invented by any one man or group of men, but evolved by almost imperceptible stages, somewhat as language is believed by some to have developed from meaningless cries. Somewhere, somehow, human beings may have drifted into the habit of using numbers without knowing what they were doing. Nonetheless, the numbers 1, 2, 3 . . . exhibit some of the marks of sudden inspiration and conscious invention. The most significant of these are again connected with the universality and invariance of numbers. Although nobody knows whether such a thing ever happened, it is tempting to imagine that some nameless genius quite suddenly perceived that a man and woman, a stone and a slingshot, a dream and a sunset, and in fact any couple of things whatever, are all alike in one respect and only in one: their "twoness." From there to the conception of the number two itself was a gigantic stride, but some man must have taken it centuries before the King reviewed his captives.

Lest all this still seem too easy, let us accept the number two as the commonplace it appears to be, and ask ourselves what two, considered as a number independently of its uses, "really is." In short we are to define the "number" two in a manner acceptable to at least some (but not all) twentieth-century mathematicians. A similar definition is to hold for any natural number.

It is not easy. Between the counting of the 1,422,000 goats and a reasonably satisfactory definition of two, there is a blank of about 5500 years in which neither mathematicians nor logicians could satisfy themselves what two is on its own merits. With the caution that finality is the last thing any instructed mathematician strives to attain in mathematics, we shall merely state the definition. Two is the class of all those classes of things which can be paired off, one-to-one, with the members of any couple of things. "Class" is to be understood intuitively as a primitive notion not further analyzed. The apparent circularity in "two" and "couple" is only accidental and can be avoided. Thus the natural "number" two is a "class"; and similarly any natural number is a class.

Without attempting an analysis of this rather recondite definition, we note that when pondered and understood, it captures what eluded the first man who observed that all such collections as a husband and his wife, a dawn and a death, a bird and a thunderstorm, have in common only their twoness. This observation, whoever made it, was the beginning of arithmetic. It was also the secret source of all the magic of numbers that insinuated itself into ancient philosophy, mediaeval number mysticism, and modern science.

We have noted one possible origin of numbers. In suggesting that numbers were invented we did great but unintentional violence to more than one respected philosophy of number, Plato's among them, and outraged the beliefs of many eminent mathematicians of the nineteenth and twentieth centuries. Historically an obvious alternative has been far more widely accepted. If numbers were not invented by human beings they may—not necessarily "must"—have been discovered. Here is the parting of the ways, where knowledge ends and opinion begins.

Some mathematicians believe that numbers were invented by human beings. Others, equally competent, believe that numbers have an independent existence of their own and are merely observed by sufficiently intelligent mortals.

The difference between the two creeds is anything but trivial. Both cannot possibly make sense. It is conceivable, however, that the question "Were numbers invented or were they discovered?" is improperly posed. It may seem as meaningless to our successors as the question "Is honesty blue or is it triangular?" seems to us. But at present—until the psychologists intervene—the question about numbers seems to make as good sense to us as some others which can be answered unambiguously. For example, "Was America discovered in 1492 or was it invented then?", or "Did Watt invent the steam engine or did he discover it?"

Even superficially these four specimen questions are of different types. Though the one about honesty has the grammatical form of a meaningful question, it is merely a nonsensical string of words. The one about America could be quickly settled, except perhaps in a metaphysical debating society, by accepted methods of evaluating historical evidence. The question of Watt and the steam engine might be resolved similarly. Then some thoughtful philosopher might remark that the eternal structure of the physical universe and the constitution of the human mind necessitated the invention of the steam engine sooner or later in the destined unfolding of history. Without laboring the point we note that a case might be made out for Watt as part inventor, part discoverer. It is even possible to make some kind of sense of the assertion that the steam engine was waiting to be discovered ages before the solar system came into being. Watt would then be a mere observer of the already existent.

The question about numbers—were they discovered or were they invented?—cannot be disposed of by any such means as suffice for the one about America. Whichever answer we favor is determined largely by our emotions. For plainly the question is unanswerable by any objective or documentary test, and yet it is not, apparently, nonsensical. In this it resembles several profounder questions concerning man's relation to the universe that have exercised philosophers, theologians, and scientists for many centuries. Those who would say numbers were discovered might agree that man is the noblest work of God. Those favoring a human origin of numbers would be inclined to retort that man almost invariably has made his gods in his own image.

It is not necessary to take sides in this age-old controversy. Our only concern here is to observe certain phases of it down the centuries, and to note how deeply men's beliefs concerning the Platonic reality of numbers—their existence as suprahuman "entities" beyond man's interference—have influenced their beliefs in other fields far distant from mathematics and perhaps of greater human value. Whether the question "Were numbers invented or were they discovered?" is answerable or unanswerable, or whether it is meaningful or improperly posed, its impact on the development of rational thought more than once has been decisive. Emotional or rational attempts to answer it continue to generate controversy, if nothing more profitable. It is the oldest and the simplest of all questions regarding the nature of mathematical truths. History gives no universally accepted answer to it; science, it is hoped, may.

Instead of trying to come at the origin of numbers by conjectural reconstructions of the history of our race, psychologists have sought the same goal by imagining the early development of the individual. Counting becomes a possibility to the future arithmetician when, as a very young baby, he falls out of his crib or bumps into a chair. For the first time in his life he then senses the "not-I." The "I" and the "not-I" are the matrix of all plurality. It may not be too fanciful to see in this shattering recognition of a hostile "not-I" the subconscious beginning of the evil associated with the number two by all number mystics from the ancient Pythagoreans to the theological numerologists of the Middle Ages. Two, the "Dyad," the "not-One" invariably is represented as unstable and bad, as deceptive indeed as a two-dollar bill. The number-wise Dante (thirteenth century), for example, argues that the Empire should be "unified" because "being one" appears as the root of "being good," and "being many" the root of "being bad." It is for this reason that Pythagoras puts "One" on the side of good, and "Many" on the side of evil. Dante might have added that Plato followed Pythagoras in this respect and that each may have been recalling the subconscious memories of his infancy. Unless the future number-mystic is also a born solipsist he will learn very early that he is not the all-powerful, all-knowing One and Eternal Monad.

Further painful encounters with tables instead of chairs may lead to the perception of "not-chair." The infant's loving parents and the not-so-loving family cat impress further distinctions on his raw and tender consciousness. But unless the infant is to be a great mathematical philosopher, he will not intuit his parents and the cat as sharing anything universal with the inanimate trinity of two chairs and a table. Indeed he will probably never discover (or invent) "3, 4, 5 . . ." by himself, but will have to be taught them by his parents. From whom did his parents learn the numbers? From their parents. And so on, back to savagery.

At this point the psychoanalysis of number becomes somewhat less sure of itself. From whom did the savage learn? His parents stopped at the number six. Did the genius of the tribe invent the "seven" he used to count the arrows his father was unable to enumerate? Or was "seven" waiting to be summoned from the realm of Eternal Being? And will it still be there when the human race is extinct, ready to be rediscovered by some future species of intelligent animals? How much of "number" is created by the human mind—or by human behavior—and how much is self-existent and only observed? It will do the practical man little good to say that only a metaphysician would ask such questions. The historical fact is that numerous impractical men not only asked these questions but struggled for centuries to answer them, and their successes and failures are responsible for much by which the practical man regulates his life in spite of his impatience with all metaphysics.

As usual in such inquiries the favored answer is an inconclusive compromise. Experience teaches the savage that number is a reliable label for distinguishing objects whether like or unlike. Once he has perceived the difference between one thing and many, the savage is compelled (by what?) to continue through "three" things to "four" things, and so on for as far as may be profitable. Only at a much later stage, when civilization is a habit, do fairly general conceptions of numbers emerge. At some intermediate stage such arithmetical facts as $4 = 2 + 2$, $4 = 1 + 1 + 1 + 1$ must have been apprehended at least intuitively. Any conception of numbers which contradicted these basic facts of arithmetic as we know it would be rejected as too clumsy for use.

Though abandoning the main question unanswered, this compromise has the double merit of leaving open two doors, one to naturalism, the other to supernaturalism. After the first step hesitation was no longer possible. The scores of mystics, philosophers, and mathematicians who chose the second door beheld the vision of number as a divine creation. Some even saw number as the power to which even the gods must bow. Those who preferred the way of naturalism found nothing superhuman. Their negative reports were largely ignored, and they themselves achieved no great popularity. The few independents who refused to enter either door and maintained their open minds had almost no support.

The next significant historical episode after the royal mace of 3500 B.C. concerns the Babylon of fifteen centuries later.

CHAPTER - - - - - - - - - - - - - - - 3

For Their Own Sake

NUMBER for the ancient Egyptians remained a strictly practical concern all through the recurrent flow and ebb of their fluctuating civilization. Consequently Egyptian arithmetic made but little progress, crystallizing in a singularly uncouth form about 1700 B.C.—about the time Stonehenge was completed. There is no paradox in this. In mathematics as elsewhere close attention to immediately useful ends is not always the most effective way of being practical. Curiosity about numbers for their own sake and interest in things of no evident value were necessary before mathematics could advance. Astronomy and the physical sciences then began to accelerate, and with them, technology.

With the progress of so-called pure mathematics even in ancient times the scope and power of materially useful calculation increased enormously. Practical problems which an Egyptian of 1700 B.C. could solve only to a first crude approximation were completely solvable to any required degree of accuracy by the Greek methods of fourteen centuries later. For example, the amount of grain that could be stored in an Egyptian granary shaped like a modern silo was computed by a wastefully inaccurate rule inferred from cut-and-try experience. The Greek method, deduced by pure geometry, was capable of stating the amount to within a handful.

The Greeks have been mentioned ahead of their proper time because until about 1930 it was commonly believed that they were the first people to develop the sciences of number and space for other than obviously practical ends. Through the accurate interpretation of dozens of the Babylonian baked clay tablets it appears now that the Greeks had forerunners in the pursuit of the immediately useless. Only one item of all the marvelous things the arithmeticians of the Euphrates valley did while the Greek tribes were still roaming over Asia Minor as half-civilized nomads is important for our purpose. But it will be of interest to glance for a moment in passing at what the Babylonians accomplished toward the creation (or discovery?) of mathematics.

Perhaps the most remarkable feature of this work is the vanishing of the best of it from human remembrance for thirty-five centuries or more. Certainly the Greeks must have overlooked the arithmetic and algebra of the Babylonians, or their own—the rudimentary algebra disguised in elementary plane geometry—would have been less awkward than it was. Except in numerology the early Greeks did not excel in either the theory or the practice of numbers.

The initial impulse for Babylonian arithmetic appears to have been imparted by the Sumerians, a highly gifted non-Semitic people inhabiting the fertile lands at the north end of the Persian Gulf. Among their other outstanding contributions to civilization the Sumerians invented a semi-pictorial script which developed into the cuneiform writing of the Babylonians. This proved adequate for the preservation and transmission of arithmetic. By 2500 B.C. the merchants of Sumeria were familiar with the applications of arithmetic to weights and measures, interest on loans, and what we call commercial paper. Their efficient use of numbers suggests a long previous development, possibly as much as a thousand years. About 2000 B.C. the Sumerians were absorbed by the Semitic Babylonians, and the golden age of Babylonian mathematics began. It lasted all of eight centuries.

Babylonian counting was based on the sexagesimal system of numeration (by 60's) with a slight admixture of decimal counting (by 10's). The base 60 survives in our measurement of time, also in our degrees, minutes, and seconds for angles. Both whole numbers and sexagesimal fractions were represented by cuneiform characters in a place system of numeration (to the base 60), substantially as our own numbers and decimal fractions are written (to the base 10) in the simpler symbols 0, 1, 2, . . . 9. At some unknown date, but probably late in the climactic period, a character corresponding to our zero was introduced. This alone was an advance of the first magnitude.

Although it is of interest for the history of mathematics rather than for our own narrower purpose, we may note in passing that the work in arithmetic led quite naturally to rules for the numerical solution of quadratic, cubic, and biquadratic equations. Though the Babylonian algebraists could not completely solve an equation of any of these types given at random, as is done in high-school algebra today, they made great progress. Some historians of mathematics rate this Babylonian algebra of 2000-1200 B.C. as superior to any other produced before the sixteenth century of our era. The work in geometry and mensuration is almost as astonishing. Though the results are for the most part correct, there is no vestige of proof. This absence of proof is the point of critical interest in the historical development of reason and philosophy.

One apparently trivial but historically important detail of the arithmetic will reappear when we reach Plato. Large numbers, and in particular one, seem to have attracted the Babylonians. The number in question is 12,960,000 or the fourth power of 60 ($60 \times 60 \times 60 \times 60$). In the sexagesimal system this number would be the "ten thousand" (fourth power of the base); our "ten thousand" is 10,000 ($10 \times 10 \times 10 \times 10$). Theirs may have been used, as the Greeks used ours on occasion, to signify a vaguely large number. Plato's use of the Babylonian "ten thousand," as will be seen later, was incomparably more imaginative.

One source of the curious things this number has suggested to magicians and others is the comparative largeness of the number of its divisors. Including 1 and itself, the Babylonian "ten thousand" (12,960,000) has 225 divisors; our "ten thousand" (10,000) has only a paltry 25. If this strong hint for a metaphysics of an eternally recurring universe is not enough, we need only observe that 225, the total number of divisors of the fourth power of 60, is 9 times 25, and that 9 is 3 times the omnipresent and ever sacred 3. And if this still is insufficient, we note that the fourth power of 6 ($6 \times 6 \times 6 \times 6$, or 1296) has exactly the same number (25) of divisors as the fourth power of 10. But the fourth power of 10 is the "ten thousand" of the Greeks and ourselves, while 12,960,000, the "ten thousand" of the Babylonians, is equal to the fourth power of 6 multiplied by the fourth power of 10. Surely there must be some profound cosmic truth concealed in these mysterious harmonies of numbers? Whether or not there is, vast philosophies of man and the universe have issued from conjugations of numbers less prolific than these. It will be refreshing now to look rather carefully at something utterly useless.

In American colonial days, and well into the nineteenth century, the scholars in the top class in arithmetic wrestled with such brain-twisters as the following. "A parcel of land containing 1000 square feet is the sum of two squares. Two-thirds of the number of feet in the side of one square exceeds by 10 the number of feet in the side of the other square. Calculate the sides of the squares." Algebra gives two answers: the sides of the squares are 10 and 30 feet, or $-270/13$ and $-310/13$ feet. Arithmetic more sensibly gives only the first.

An unjustly forgotten American classic of "mental arithmetic" bristled with such terrors. The hardy lads who solved these problems in their heads (they could get only the first answer; the second is nonsense) must have been of a tougher breed than the pallid weaklings who later squandered algebra and pencil and paper to find both answers. There were even more exasperating problems than the specimen exhibited—like the teaser about the escaped prisoner of war who could carry only enough food and water for two days and had to cross a waterless desert a hundred miles wide in marches totaling ten miles a day. But however diverse the problems, they all had four features in common. They could be solved by common arithmetic by anyone sufficiently good at arithmetic; they could be solved much more easily by anyone who was only moderately skilled in elementary algebra; they were elaborately artificial and of no practical use whatever; the pupils with strong heads liked them.

The last two are the items of significance here. In a modern progressive school the problems in arithmetic have a wisely practical air, presenting as they so frequently do (with pictures) interesting and important facts about the local school board, the Grand Central Terminal in New York, and the activities of the city fathers. They also can be done in the head—by almost anybody. Rather to the bewilderment of some of the more progressive teachers about ten per cent of an average class thoroughly dislikes these speciously practical problems, and even clamors occasionally for something that will make a normal boy or girl think. The ancient Babylonians seem to have had a similar experience.

By 2000 B.C., and possibly even as early as 2500 B.C., the Babylonians had developed an arithmetic powerful enough to take care of their daily activities in trade, in agriculture, in building, in canal digging, in astrology, and in astronomy. They then turned to unpractical mathematics, proposing and solving numerous problems that not even the most reckless economic interpreter of history could claim were of the slightest use to anyone. The problem about the parcel of land is a mild specimen of what they did in this direction. It is from a mathematical tablet of about 2000 B.C. Its sham air of practicality would not deceive a surveyor for a moment. If anyone wanted to know the sides of the squares, he would take no such measurements as those stated in the problem—unless he were either insane or preternaturally stupid. The problem is as artificial as an anagram, and the only possible gain from solving it is the satisfaction of having exercised one's brains.

The Babylonian mathematician who posed and solved this problem permitted himself to use algebra. Until almost the last step he followed exactly the same method that many pupils in the first course in algebra follow today. As negative numbers were not yet fully known, he missed the second answer and gave only the first. The second, had he been just a step more advanced, might have puzzled him exceedingly. What does a negative length look like? Many a beginner in algebra has asked that question, only to be assured that since figures cannot lie they always make sense if properly handled; hence negative lengths are to be discarded.

It was only after candid people stopped throwing awkward things away, and made an honest effort to understand what they were doing with numbers (or what numbers were doing with them?) that arithmetic began to develop fully and freely. But this was not until many centuries after Babylon had disappeared in thousands of tons of dissolved brick, and "the glory that was Greece"—its mathematics—was about to be rediscovered by awakening Europeans. Even then centuries were to elapse before negative numbers were manipulated with justified confidence, and it was only in the nineteenth century that complete understanding came. But the capital question for philosophy, "Were negative numbers invented, or were they discovered?" remained unanswerable.

Though the Babylonian algebraist found only the sensible answer to his problem, he had shown anyone with eyes to see something that was to be of infinitely greater importance than the missing answer for the futures of science, mathematics, and philosophy. By his own conduct, he had demonstrated that numbers are attractive in themselves and reward study for their own sake. Curiosity may be idle if allowed its own way too long; but without it, little of even the lowest practical value has been achieved. This is the teaching of Greek mathematics and science. The honor for having taught the world this fact must be shared with the Babylonians.

CHAPTER - - - - - - - - - - - - - - 4

The Decisive Century

THE seventeenth century of the Christian era has many impressive claims to the title of "the great century" for modern science and mathematics. That was the century in which Galileo (1564-1642) and Newton (1642-1727) first exercised the full power of our present scientific method which combines mathematics with observation and experiment. Neither believed, as do some of the twentieth-century theoretical physicists, that pure reason—mathematics—alone can reveal all the fundamental laws of the physical universe.

To disbelievers in number magic it seems unlikely that without the Galilean-Newtonian method of exploring nature the industrial revolution of the late eighteenth and nineteenth centuries would ever have happened. The transformation of human life induced by the application of modern science to practical affairs is too familiar a story to need retelling here. It has been recalled merely to provide an adequate comparison for the revolutionary progress of civilization in another great century, the sixth B.C.

In that century two Greeks, Thales and Pythagoras, the first immortals of the exact sciences, started definitely toward the science and mathematics which made possible the work of Galileo and Newton. In other respects also it was a memorable century for what was to be the future of western civilization. If sharp turning points in history are more than fictions of inventive historians, the sixth century B.C. was one of them. Scientific, mathematical, and religious thought took new directions in that century, veering partly away from the authority of age-old traditions to question nature and human aspirations directly. After Thales and Pythagoras had lived it was no longer mandatory to approach the gods through the mediation of priests. Side by side with the crudest superstitions, rational speculation regarding the physical universe and man's relation to it flourished, possibly as never before and as but seldom since. But not even the most daring men of the age could completely throw off the burden of the past in their short lives. To the lasting confusion of reason, the boldest of them all, Pythagoras, transmitted to future generations the number magic of the East along with his own epochal contributions to experimental science and mathematics.

Before observing what numbers did for the thought of Pythagoras and his successors down to the present, it will be interesting to notice the intellectual climate in which he and his immediate predecessor Thales prospered. They were not alone in their diversion of human thought into new channels.

Nearly three centuries before Thales (624?-546 B.C.) and Pythagoras (569?-500? B.C.) were born, Homer (about the ninth century B.C.) gave the common man of Greece his immortal gods in an all too human shape, which was to remain popular and orthodox for a thousand years. Beyond two of the timeless masterpieces of all epic poetry, he gave the uncommon man nothing. Homer's meddlesome gods and goddesses were as incongruous when Thales devised the first proofs in geometry as were Milton's cannonading angels when Newton applied the differential calculus to celestial mechanics.

By the time Thales arrived, something more dignified than the Homeric ideal of the Father of Gods and Men as a lecherous old despot was available for those who had grown rather tired of degrading mythologies. The Iranian teacher Zoroaster had promulgated a more civilized conception of religion in which the ethical element was almost as prominent as the supernatural. Some of this may have influenced Pythagoras, whose own religious teachings, except those directly inspired by oriental fantasies or the nonsense of numbers, were clean of superstition.

The long war between polytheism and monotheism was well under way when Thales began his geometry. In the century before he was born a notable tetrad of Hebrew prophets —Amos, Hosea, Micah, Isaiah—whose words survive in our Old Testament, had urged the Israelites and others to reject a plurality of gods in favor of one. The Pentateuch also was probably completed by this time. It provided the Hebrews with an inspired history and a strict moral code which the orthodox strove for centuries to respect and obey. Details of this code affect the lives of practicing Christians today.

Tradition makes Thales an indefatigable traveler. For the moment we are interested only in the memorable events which were happening beyond his range and of which he probably heard at the most only rumors. While he was going about his business in Egypt, Babylonia, and elsewhere, a triad of Hebrew prophets were vigorously and incisively administering their jealous Jehovah's estate on earth. Zephaniah whipped Judah into line by alternate warnings and threats; Nahum announced that the recent overthrow of Nineveh was Jehovah's work; while Habakkuk carried on a spirited controversy with Jehovah over the oppression of the faithful. Not all of these matters are as fresh today as they were when Thales went trading up and down Asia Minor and gathering the seeds that, some two and a half centuries later, were to blossom out at last in Plato's transcendental arithmetic of Ideal Numbers. Yet they all contributed significantly to the religion which Europeans were finally to accept in preference to all others. Thus the mathematical, scientific, philosophical, and religious ideas which have guided western civilization were already in embryo in the sixth century before our era.

Asiatic culture, too, took on some of its permanent set in that amazing century. Confucius invited the Chinese to practice one philosophy of life, Lao-tse another which was to survive as Taoism. The Indians were offered, and accepted, Buddhism and Jainism in the teachings of Gautama and Mahavira.

Neither the Chinese nor the Indians had yet made any contribution of outstanding significance to the science of numbers. One peculiarity of Indian number lore, however, does seem to have affected the metaphysical speculations of the earlier Greeks. The Indians reveled in large numbers, especially in their pantheon and their mythical chronology. Like the Egyptians they had overcome the primitive reluctance to count boldly. A step or two farther along the same road, and they might well have imagined the infinitely great.

To conclude these citations of memorable names, three

from the time of Thales may bring the atmosphere of his age nearer our own. While Thales was shaping the beginnings of deductive reasoning in the strict mathematical sense, a triad of Hebrew prophets—Ezekiel, Haggai, Zachariah—kept exhorting Israel to cease backsliding lest it be smitten by the wrath of Jehovah, and to complete the building of Solomon's Temple in Jerusalem. They also foretold the advent of a prince of peace who would deliver the world from war and other evils.

It may be only a baseless fancy, but in looking back it seems that our civilization narrowly missed turning East instead of West in that critical sixth century B.C. For when Thales died, Gautama, the Buddha—"The Enlightened One" —was about fifteen years old. Overlapping Pythagoras, Gautama outlived Thales by some sixty-five years.

Thales and the Buddha never met. There is a tradition, however, probably without foundation in fact, that Pythagoras on his legendary travels encountered the Buddha. If they had met, what ideas could these two men, universally counted among the most influential teachers of all time, have exchanged?

The Buddha, with his insistence on right thinking as one step on his eightfold path to perfection, would have delighted in the Greek's attempt to state precisely what just one kind of right thinking is. But there is no evidence that the Buddha ever heard of the mathematics Pythagoras was revealing with all the zeal of a first pioneer exploring a newly discovered continent. Pythagoras for his part might have learned even more than he already knew of the transmigration of souls and the mysteries of successive reincarnations. For wherever he acquired those enervating oriental beliefs, which today hold millions of untouchables in resigned degradation, Pythagoras held them as tenaciously as any Indian fakir. They and his passion for numbers bred a fantastic metaphysics that was to transmigrate through creed after creed until, purified at last of all sensible taint, it sank to its Nirvana in the refined number magic of twentieth-century physics.

Had Pythagoras and the Buddha met, it is just possible that the world might have been spared the three centuries of experimental science that followed Galileo and Newton. Indeed this shortcut to an understanding of the physical universe might well have started immediately after the meeting, and Plato, not Newton, would have stated the law of universal gravitation. Half a generation later, Einstein would have been incarnated in the body of Aristotle.

Unfortunately for this consummation of knowledge and wisdom, Pythagoras himself had dabbled in scientific experiment and had been lifted out of his capacious ego by the experience. Science, mathematics, and philosophy hesitantly turned West, not East.

A consistent believer in modern number magic would be forced to conclude that this turn to the West delayed the industrial revolution till the late eighteenth century A.D. A turn to the East would have precipitated it in the third century B.C., and World War II could have been fought about the year one of our era. What state the world might now be in, not even the most expert numerologist can figure.

CHAPTER - - - - - - - - - - - - - - 5

A Difference of Opinion

WHEN asked by his grateful fellow citizens what reward he would like for his services to them and their city, Thales replied, "Credit for my discoveries." He is the first man on record to guess that intangibles of the mind may outlast material wealth.

It was a shrewd guess. Rich King Croesus specialized in gold. His moderately well-to-do friend, the astute Thales, preferred ideas. His eyes were fixed on immortality. If Croesus, reputedly the wealthiest man of antiquity, contributed anything of value to civilization beyond the simile "as rich as Croesus," it has long since been forgotten. And though Croesus as a mere name is probably more widely remembered than Thales, it is the latter who continues to live. One of his achievements alone earned him the immortality he craved. Deductive reasoning as used in geometry is traditionally ascribed to Thales. He only glimpsed what Pythagoras and his disciples were to develop into the sure beginnings of mathematics as understood today; yet he was the first man on record to envision its possibility.

As will appear later there may be some grounds for crediting the ancient Egyptians with deductive reasoning in geometry. But beyond the highly ambiguous testimony of one man, no evidence to this effect has been discovered. By Greek tradition and history, Thales in the sixth century B.C. was first.

The relation of deductive reasoning to all mathematics and science is so important for the sequel that a little may be said about it here, before we proceed to Thales himself. To expose the heart of the matter as nakedly as possible, without deductive reasoning, mathematics as understood by professional mathematicians does not exist. This categorical statement of fact usually enrages those romanticists who delight in finding prodigious feats of mathematical genius in everything from the inventories of a mummified Egyptian steward to the zigzag lightning on a Zuñi cooking pot. Nobody would deny that such things may have preceded arithmetic and geometry, or that they may have suggested mathematics to men capable of deliberate and purposeful abstraction. But to mistake them for mathematics is to confuse all reasoning in a roseate haze, in which the mythologies of savages are indistinguishable from the gravitation of Newton and the space-time of Einstein. Failure to distinguish between what mathematicians call mathematics, and the semi-empiricism that preceded this mathematics but was sometimes mistaken for it, misled numerous philosophers from the ancient Greeks to Kant in the eighteenth century. This will be resumed in the proper place.

"Deductive reasoning" may be replaced here by the shorter and equally descriptive word "proof." Two details will suffice. Proof in mathematics proceeds from definite assumptions explicitly stated. The assumptions are variously called postulates or less frequently axioms. In ancient times it was believed that the postulates of mathematics are necessary truths inherent in "the nature of things" and indispensable to any consistent (not self-contradictory) account of "number" and "space." This belief in the eternal necessity of the postulates of, say, elementary geometry and arithmetic, persisted into the nineteenth century. Then it was gradually perceived that the postulates underlying mathematics are not necessary truths in the sense described, but are conventions agreed upon by mathematicians. In particular, the postulates of geometry are of human origin. They were not imposed on human beings by "the nature of things," or by any other extra-human agency. This very inadequate summary of a two-thousand years' dispute must suffice here; it will be elaborated later.

The second detail to be kept in mind concerns the process by which mathematical conclusions are derived from the postulates. This is called deduction. The postulates are accepted as "true" without further argument. Any statement implied by the postulates is then true, merely by definition. The task of mathematics is to find what statements are implied by the postulates.

It is sufficient here to note that only a system of reasoning agreed upon by mathematicians is to be used. This system is called formal logic. Since its origin in ancient Greece it has been vastly extended by mathematicians till today the classical logic of Aristotle is only a comparatively unimportant detail in the formal—or mathematical—logic habitually used by mathematicians. Like the postulates on which it operates, the logic of mathematicians is a matter of common agreement among mathematicians. It is not imposed upon them by fate or eternal necessity. This also needs elaboration; but for the moment it is enough.

We are not concerned with the reasons why mathematicians prefer some systems of postulates in the various subjects to others that may easily be imagined, or why they use certain types of reasoning in preference to others. Roughly, it appears historically that the geometers of remote antiquity drifted into certain profitable habits of thought suggested by reflecting on practical experience. Before they consciously knew what they were doing, they were reasoning deductively. The conclusions of their reasoning always seemed consistent with one another.

From this some of the more philosophical mathematicians drew the grandest and least logical conclusion of all: logic is a necessity, an eternal fate, imposed upon the human mind from without. It is not an invention of men but a timeless gift to mankind from the immortal gods. In one form or another this belief has persisted for well over two thousand years. Doubts as to its usefulness are quite recent.

The foregoing account may have overstressed the claims of one school of philosophy on these basic questions at the expense of its rival. Did Thales (or some other man) invent deductive reasoning or did he discover it? The question is of the same kind as the one about numbers: Were numbers invented, or were they discovered? There is no need to repeat for deductive reasoning what was said about numbers. Each may choose the answer which best accords with his temperament. Great minds have disagreed. For ourselves, we shall be satisfied as we proceed to see how this irreconcilable difference of opinion came about.

What becomes of the Egyptian and Babylonian work in numbers and the rest under the narrow conception of mathematics described above? As neither people ever proved anything—so far as has yet been discovered—their contribution was not mathematics. Nobody is compelled to accept so disconcerting and so ungracious a conclusion, and few will. Perhaps it is neither necessary nor profitable in the usual historical accounts to make a sharp distinction between what is to be called mathematics and what is to be denied that high title. The insistence on proof as the criterion is a fairly recent demand. If strictly enforced it rejects too much of what our predecessors called mathematics and seriously encroaches on our own.

A compromise would admit whatever a majority of the competent mathematicians of a particular epoch accepted as proved, whether it withstood the criticism of later mathematicians, or whether it was shown to be unsound or incomplete. This would make the test a recognition that there is such a thing as proof. Those who tried to prove their results would then be mathematicians, the others, empiricists.

The distinction is familiar enough to editors of mathematical periodicals who must decide whether the contributions submitted to them for publication are mathematics or something else. To take an example from arithmetic, an assiduous calculator observes after forty years of grueling labor that 8 and 9 are the only numbers less than a billion billion which differ by 1 and are both exact powers ($8 = 2^3$, $9 = 3^2$). Having worn out several calculating machines and no inconsiderable part of his nervous system, the would-be mathematician decides to call it a day and guess the rest. So he writes to the editor of his favorite mathematical journal, announcing his conjecture: "The only exact powers that differ by 1 are 8 and 9." "You may be right," the editor replies, "but how do you prove it? Hoping to hear from you in the near future, I return your manuscript." He is still hoping.

CHAPTER - - - - - - - - - - - - - - 6

Wisdom as a Profession

IN THE career of Thales we observe hints of a new leisure class and the rise of a novel cult, that of the professional wise man. Unless the philosophers and mathematicians of early Greece had been relieved of manual labor, it is unlikely that they would have contributed much to either philosophy or mathematics.

The exceptional man who lived well without doing anything the common man would call work was no rarity by the sixth century B.C. Indeed long before then there were several thousand of him at one time in Egypt alone. These blest mortals swarmed like drones about the temples and the court of the King, and made their livings by transmitting the will of the gods to the King and the common man.

Thales and his successors in the profession of wisdom made no pretense of giving the common man anything of value to him as had the priests. The wise men of the new order stood erect without support from the gods and squandered scarcely a thought on the slaves who made it possible for them to eat regularly. Some of these sturdy thinkers were independently well off, others graciously accepted a living from wealthy patrons.

The most remarkable feature of this alliance between material wealth and pure thought is absence of the profit motive. The priests had promised the kings eternal rewards, and some had even hinted that the slaves would be generously recompensed after they were dead. The intellectuals promised nobody anything. They may have been too honest to assume obligations they felt themselves unable to discharge. Nor did they ever dream that centuries after they had ceased to exist their useless work would have helped to free the slave from brute labor and the King from servile superstition.

The first to glimpse mathematics as it is understood today, Thales was also the first layman to make wisdom a profession. When asked by his admirers what they should call him, he chose the title of "sophos," a wise one. And wise he was, sometimes too wise in fact for the prosperity of his neighbors. He was not a perfect specimen of the professional wise man who was to follow so shortly after him; for he made his own living at what might have passed in his day as honest work.

The son of a Greek father, Thales was born in Miletus, Ionia, in the seventh century B.C. His birth date is given as 640 or 624, the latter being the more generally accepted, and he was still living in 548. His father's name was Examius, his mother's, Cleobuline. This is all that is known about them, except a legend that Cleobuline was of Phoenician extraction. There may be some fact behind this, though "Cleobuline" is said to be a good Greek name. A drop or two of Phoenician blood in his veins would account for much in Thales' career. For there was an ancient tradition, current among those who did not like the Greeks very well, that the Phoenicians, the sharpest traders in history, taught the Greeks how to trade in everything from clipped coins to wooden horses. One of Thales' masterpieces in this difficult field might have been transposed intact from the twentieth century.

Among his other accomplishments Thales was an entrepreneur. Foreseeing a bumper crop of olives one spring in Miletus and Chios, Thales unobtrusively cornered the oil presses. When late summer came, and the olives began to drop, the growers paid Thales what he persuaded them was not an exorbitant rental for the use of the presses. The crops were saved. In this way, perhaps, Thales financed his protracted education in the temples and market places of Egypt and Babylonia. The Milesians and Chians also learned something from the transaction. Thales did not return to that part of the world for many years.

Olive oil was therefore indirectly responsible for some of the Thalesian philosophy. Of course the future philosopher might have made his way to Babylon and Thebes without oil to barter along the route; but his progress would have been less carefree. The oil and salt trades financed more than one prying Greek on his tour through the East; even Plato is said to have peddled oil in Egypt.

The high financial skill exhibited by Thales in his handling of oil has more than a sentimental value for the histories of mathematics and philosophy. It offers a perfect example of deductive reasoning in action.

"I want to study in Babylon and Thebes," was Thales' first postulate. "I have not money enough to get to either place unless I walk all the way and live off the country," was his second. "I don't want to do either," was his third. He then stated a lemma: "If I can induce somebody to give me a handsome sum of money, or its equivalent, I can travel all over the East like a gentleman and study what I like in comfort."

Considering this lemma too obvious to require a formal demonstration, Thales stated a definition: "Olive oil is equivalent to money." This was an example of what the philosopher Kant was to call a synthetically true proposition. It was a matter of observation which could be verified by appeal to actual experience in the real world. Every reputable scientific theory contains such propositions; otherwise it has no relation to reality.

Having inserted the indispensable link to connect his pure thought with the concrete world, Thales returned to the abstract and proceeded rapidly with his unbreakable chain of deductions. A condensed summary of his principal definitions and main theorems will be enough here. "Olive oil is equivalent to money. The oil is squeezed from ripe olives by presses owned by the growers. The feckless growers are as hard up this spring as they always are between crops. Therefore they will part with their presses for about one per cent of their value. To squeeze the growers dry of their last drop of oil (which is equivalent to money) next autumn, it is necessary and sufficient that I own all the presses. To get the presses for what I can pay, it is necessary that I pledge each seller to secrecy, after convincing him that he is smarter than his neighbors, whose presses he will expect to borrow for nothing. Except for their free borrowing and lending, the growers are all rugged individualists with no concern for one another's welfare. Therefore I shall travel like a king, and study or not as I please, whatever and wherever I please." History and tradition record that Thales spent several years in Egypt and Mesopotamia studying arithmetic, geometry, and philosophy.

Neither the Egyptians nor the Babylonians succeeded in teaching him anything about finance.

It should be remembered that this pioneer of mathematics and philosophy is known to us only through legends and references to his teachings by later mathematicians and philosophers. No contemporary record of his life or sayings has survived; and it is possible that we have a ludicrously erroneous picture of Thales the man. But as anyone in science or mathematics knows who has read (or written) the biography of some recently deceased colleague, the official life of a notable man frequently leaves a flatter and falser impression of his life and character than does the three-dimensional personality of the uncertified anecdotes, a good many of which may be quite untrue. A devoted or satirical pupil of some famous master can sometimes immortalize the great man and preserve him for future generations as he really is, like a beetle in amber, in a single phrase. So it may be with Thales and Pythagoras, neither of whom boasts a biography that would satisfy either a pedant or a critical scholar. If nothing else, these unverifiable legends embalming the teachers of antiquity show us what common mortals thought of them. And that surely is as important for understanding as knowing where and when these great men lectured, and about what, on each documented occasion.

Another of the classic legends concerning Thales is of particular interest here, as it shows that deductive reasoning is not exclusively a prerogative of human beings. For years at a time Thales followed the highly lucrative salt trade, transporting his precious cargoes by mule pack train. Now mules, as anyone who has been privileged to collaborate with them will testify, are among the most intelligent creatures the devil ever made. One of Thales' mules was easily in the genius class. While fording a stream one day, this super-mule slipped on a rock and fell into the water, thoroughly dousing its load of salt. At the next ford it lay down and rolled in the water. Evidently it had observed that the first fall lightened its burden. The trick was repeated till the salt had all dissolved away. Then that most logical mule stopped rolling. To cure it of these extremely intelligent misdeeds, Thales substituted a load of dry rags and dusty sponges for the salt. Thereafter the mule rolled but once.

Obviously this history exhibits all the basic elements of both induction—inference and generalization from repeated experience—and deduction. A man who could think of the remedy Thales prescribed to circumvent the mule's intelligence was an applied scientist ripe for the invention of the deductive method and the creation of mathematics. What the mule might have invented had he been endowed with intelligible speech is beyond human imagining.

All of Thales' earlier assays in reasoning have the same shrewd practical twist. They also have something much more useful for a mathematician, an ability to turn the obvious about and to see in it things which are not evident to casual inspection. The quality of his thought was different from that of his contemporaries. It must have been, or he would not invariably have bested them when they were foolhardy enough to challenge him to a duel of wits. There was the celebrated encounter with Solon (639?-569 B.C.), for instance, in which Thales proved himself a sharper lawyer than the official legislator-in-chief of Greece.

By choice and sound logic Thales remained a bachelor all his life. Meddlesome as was his wont, Solon took it upon himself as a civic duty to lecture Thales publicly for not acquiring a wife and contributing to the defense of the State his just quota of infantry. Thales meekly accepted the rebuke and promised to think it over. He did. Some days later Solon was informed in the presence of Thales that his son had been killed. Thales had contrived that Solon be so informed. The great legislator forgot all about the State. It was enough. Thales confessed that the report was a strategic lie. "You see," he pointed out, "you can't take it, and yet you want me to. Is that being consistent?"

Two of the innumerable stories his admirers loved to repeat about Thales have a simple and direct message for our own chaotic times. In his day there was incessant brawling between rival gangsters and their armed mobs—to borrow the phrases popularized in 1940 by President Roosevelt. The unit of disorder was the City State, the political entity that figures so conspicuously and sometimes so gloriously in Greek history. Five Ionian City States suddenly stopped fighting among themselves when Thales remarked that a federation of all would be secure against aggression from without.

The second incident reveals Thales as a resourceful engineer in addition to his other professions. The technical skill for his most spectacular feat of engineering could have been acquired in Mesopotamia or Egypt, and doubtless was. King Croesus, an admirer and one-time patron of Thales, wished to get his army across the river Halys. Pontoon bridges had not yet been invented and there was no time to build a permanent bridge. Croesus called Thales into consultation. The future philosopher and mathematician solved the problem at a glance. Under his direction a canal was excavated to divert the river into a temporary channel. When the army had passed over dryshod to resume its pursuit of the enemy, Thales had everything put back in its natural order, to avoid offending the river gods respected by Croesus.

Some years later Thales was entertaining his old friends with a modest account of the wonders he had seen in the East. A doubter in the audience at last reached the limit of his credulity. "Was there by chance just one thing you did not see in your travels?" he sneered. Thales pondered the question. "Yes," he admitted, "just one." "And what may that unique thing have been?" the doubter scoffed. "An aged tyrant," Thales replied.

Thales may have been remembering the time he helped Croesus on his way over the Halys to pursue the enemy. For Cyrus King of Persia, having routed Croesus King of Lydia in battle, draped him in chains. It is not known where or how Croesus died, or how old he was when Cyrus had finished with him.

The feat that made Thales the wisest man in all the world to his Greek contemporaries should be credited to the Babylonians, or possibly to the Egyptians, rather than to him. According to one version of the story, on the twenty-eighth of May, 585 B.C., there was a total eclipse of the sun. The Medes and the Lydians at the time were in the sixth year of a stubborn war. As Herodotus puts it, they suddenly found themselves fighting "a night battle." Terrified out of their wits by this ominous hint from above, the warring hosts abruptly desisted from killing one another. Shocked and frightened to the roots of their unscientific souls, they immediately concluded a peace, later certified and sealed by a double marriage between their respective reigning families.

The Medes and the Lydians all vanished from the earth centuries ago; and it matters no more to anyone how many of them fell in battle or how many stood their spears behind the door for the last time and died in bed. What gives the "night battle" its critical importance in the history of thought is the fact that Thales had predicted the eclipse. That prediction inspired the more meditative Greeks to a belief in the repetitive orderliness of nature and prepared them for the revelation of Pythagoras that numbers rule the universe.

A prediction with the degree of precision commonplace today, giving the time to a second and specifying exactly at what places the eclipse will be visible, was out of the question in the sixth century B.C. Probably the best Thales could do was to predict the year and less precisely the locality of the eclipse, say 585 B.C. and certain parts of Asia Minor. But this was ample to establish his reputation as a wise man. He had definitely predicted an eclipse within certain limits, and when it happened at the dramatic climax of the battle, his astounded countrymen gave him more credit than even he might have claimed. They eagerly granted his request that they call him sophos. They would not have believed him had he told them that the conjunction of the eclipse and the battle was less than one part in billions of billions of billions science and the rest coincidence.

There is considerable historical interest in knowing where Thales learned to predict eclipses. Obviously he could not have sat down and in his one lifetime philosophized the technique out of his own head. Centuries of patient observation had given the astrologers and astronomers of Mesopotamia the necessary knowledge of facts; and there is almost no doubt that Thales learned from them, either at first hand or through the Egyptians. As early as the eighth century B.C., those ancients were aware of the recurrent cycles of both solar and lunar eclipses. Our modern precision in these matters is possible only because Newton's theory of universal gravitation (1687) provided a basis for a remarkably accurate celestial mechanics. About twenty-three centuries separate Thales and Newton. Without Thales or his equivalent somewhere in the interval, Newton might have died an indifferent farmer.

The most lasting consequence of the eclipse is what it did for the Greek mind. Thales was the first of the wise men of Greece; his critic Solon was another of the immortal seven. "Wisdom" for the successors of Thales included what there then was of science, engineering, technology, arithmetic, geometry, and philosophy, the last in the usual sense current today. "Philosophy" for the Greek philosophers was never confined to high thinking about low living, but referred to all knowledge a philosopher was capable of grasping.

The distinction between philosophy and science came much later, when the scientific method of Galileo and Newton had so vastly increased the stock of verifiable knowledge that "natural philosophy," our physical science and mathematical astronomy, abandoned its venerable parent, and for three centuries went its own way. Lured back at last by the ancient magic of numbers, natural philosophy in the twentieth century seemed about to return to its childhood home in the sixth century B.C. There the shadowy figure of Pythagoras loomed up through the mists of time, ready to welcome the prodigal with a forgiving smile.

We may dispose here of Thales' "philosophy" in the current technical sense before considering his decisive contribution to the development of mathematics. Rather strangely for so practical a mind, Thales attempted to dissolve the universe in a single tremendous generalization. "Everything is water," he announced to his somewhat startled compatriots. Moreover, he meant exactly what he said: take anything you like apart as far as you can, and you will find yourself with nothing but water on your hands.

This was the first of the all-inclusive generalizations which the Greek philosophers and others were to offer a bewildered and incredulous mankind as ultimate summations of everything in space, time, and eternity. Admirers of Thales can only trust that he himself did not take his philosophy as seriously as did those of his Greek successors who thought it necessary to refute his grand generalization at great length.

The Babylonian origin of Thales' metaphysical water is evident. The wet fluidity of "everything" was not an utterly preposterous speculation to a people who persisted in building their cities of sun-baked mud, on a plain as flat as a floor, between two large rivers that overflowed their banks every other year. "Everything is water" sounds more like the exasperated exclamation of disgust of some Babylonian housewife than the considered contribution of a philosopher to human knowledge. It has been echoed, with a difference of emphasis, for twenty-six centuries. In the nineteenth century of our era, when the steam locomotive was lord of all it tooted over, "everything" was matter and energy, or energized ether. In the early twentieth century of humming dynamos and clattering telegraph keys, everything was electricity. In the more intellectual 1930's, when relativity had dissipated matter, energy, ether, and electricity into formulas of space-time, everything was mathematics.

In bidding Thales the man farewell before passing on his

greatest work, we remember one of the most human of all stories that have outlived the centuries. Gazing raptly at the stars one night, Thales stepped majestically into a well. Hearing the splash, followed, perhaps, almost instantly by the terrified yell "Everything is water!", a Thracian maidservant lugged the philosopher out, teasing him the while about being so concerned to see what was going on in the sky when he couldn't see what was at his very feet.

One account makes the well a mere ditch and the disrespectful servant an old woman. But Plato is our authority that the maidservant was "young and pretty." Let us believe that she was and trust that Thales, when she had fished him out, rewarded her in an appropriate manner. Of all the anonymous women who lived in the sixth century B.C., the Thracian maidservant is the one whose name I should like to know.

CHAPTER - - - - - - - - - - - - - - 7

Not Much, But Enough

NOT much in quantity, but impregnated with infinite possibilities, Thales' contribution to mathematics was enough to start the sciences of number and space on their course from the sixth century B.C. to the present time. The introduction of deductive reasoning into elementary geometry has already been mentioned and will be considered presently.

Another decisive innovation was Thales' deliberate abstraction or idealization of the data of sensory experience into pure concepts. This somewhat formidable description of a very simple process, basic for mathematics, science, and philosophy, will be discussed first. It is essential for an appreciation of much that has been called knowledge or wisdom, both ancient and modern.

A counter-example of abstraction as it appears in elementary geometry may bring out some of the main points more sharply than would an illustration of the process itself. In the 1890's an ingenious pedagogue hopefully produced a textbook of school geometry based on a new principle. His worthy aim was to make geometry not only intelligible to beginners but as easy as reading a newspaper. He succeeded in making it so hard that nobody could understand anything about it. His new principle was simply this: straight lines have a definite, measurable breadth. This, he argued, is a plain fact of everyday experience. Even the least observant must see this, he insisted, once it is pointed out to them. A "straight line" drawn with chalk on the blackboard, he observed, is sometimes as broad as a man's finger, and the cheapest magnifying glass will reveal the barely visible "straight" scratches on a window pane as irregular troughs.

The trouble began when hairsplitting adolescents clamored to know how many times a hair had to be split in two before it shredded into a bundle of true lines. Was a hair from a horse's tail as good a line as one from a girl's braid? And so on, and on, till the hapless author of the new geometry was driven within a line of lunacy. Obstinate to the last, when he finally stepped over the line into the asylum, the deluded man shouted that the lines of the geometers were all in their heads, and he dared any mathematician in the world to contradict him. In that defiant challenge to orthodoxy the new geometer had stated a fact that no mathematician (except possibly those who are also realists in the Platonic sense) would dispute. But the keepers did not let him out.

What Thales appears to have been the first to imagine was precisely the opposite of what the revolutionary pedagogue proposed to teach. The chalk marks, the scratches, the split hairs, and all the other innumerable "straight lines" of the senses, suggested a wholly insubstantial straight line, a "length without breadth," as the simplest idealization of all. This straight line of the geometers does not exist in the material universe. It is a pure abstraction, an invention of the imagination or, if one prefers, an idea in the Eternal Mind.

No longer is it possible to quibble about how broad a line is, for the phrase "the breadth of a line" has now no meaning.

This process of refining common experience and abstracting from it concepts which are not in that experience made mathematics, mechanics, and theoretical physics possible. It also inspired Plato to some of his sublimest philosophy. The geometry of lines does not refer to this or that "line," so-called, of sensory experience; it is concerned solely with certain definitions and postulates about ideas, or ingenious imaginings, useful in science and mathematics, and is valid for all lines specified by these definitions and postulates.

Geometers today know that not everything can be defined in terms of something simpler. A beginning from an irreducible minimum must be made somewhere. A start for straight lines is in such simple abstractions of sensory experience as these: "Two straight lines have one point, and only one, in common; through two points passes one, and only one, straight line." In these postulates, "point" and "straight line" are not further defined or explained. They are two of the basic, irreducible elements from which geometry is to be constructed.

It is not forbidden to think of "point" and "straight line" as common notions which every rational man imagines he understands intuitively. But any such intuitive impression is to be kept strictly in the background. It must not be obtruded into the geometry. This ban is not intended to thwart the imagination in its formulation of theorems. From the beginner of twelve in school to the master of seventy in his study, everyone doing geometry needs and uses all the intuition and imagination he has. Only when intuition and imagination have contributed their utmost are they deliberately put aside to give precedence to logic.

In theoretical astronomy and the physical sciences the procedure is the same. The earth which we inhabit and know by the testimony of our senses is not the ideal planet that figures in celestial mechanics. Ours is pocked by deep oceans and scarified by mountainous continents. The planet that figures in calculating the perturbations of the solar system is either a dimensionless particle endowed only with mass and position, or a smooth, featureless spheroid slightly flattened at its poles. Yet when the sun and all the planets of the solar system are idealized in a similar manner, the orbits of comets are computed with such accuracy that the return to perihelion of Halley's comet in 1910 after an absence of about 75 years was predicted with an error of only 3.03 days—about 1 part in 9125.

All this is so familiar to us now that we may be excused for thinking it obvious to the verge of truism. But anyone to whom the next is obvious is either a genius or a natural to whom everything is flat. Yet it too is trite in our day. It is the miracle that the ideal world of the mathematicians and theoretical scientists should sometimes foreshadow the "real" world of unforeseen experience.

To cite a famous instance, the position of a planet beyond the range of human vision is predicted (1846) by mathematical analysis applied to Newton's law of universal gravitation, and telescopes find the planet (Neptune) very approximately in the place predicted. Or, a more recent example (1927), two kinds of hydrogen molecule should exist if the algebra and theoretical physics of the modern quantum theory give a fair account of matter, and the two kinds, ortho-hydrogen and para-hydrogen, unsuspected by chemists, are found. Moreover, the amounts (¾ and ¼) in which the two kinds occur in the "hydrogen" of the chemists are as predicted. How are such predictions to be explained?

Many explanations have been proposed, too many, in fact, for any one of them to be irresistibly convincing. Only the most recent of all (1930-) need be cited here in passing, as it is germane to the magic of numbers in which it has its ancient origin. The human mind must anticipate the outcome of every scientific experiment before any experiments are performed, because it can perceive and reason consistently in only one way, and that way is mathematical, and further, the truths of mathematics are eternal. Though this may be crudely put, it is not an unfair statement of the most revolutionary scientific creed of the past three centuries. It has already been mentioned; it will recur many times and in many different shapes, from the sixth century B.C. to the present.

Some mathematicians experience a similar feeling of subjugation to necessity. Their discoveries and inventions seem to have been waiting for them in an unknown but knowable future. A rationalist might say the mathematician projects himself into an illusory time of his own devising. The future he imagines he is penetrating is his own present of abstraction and proof—the substance and spirit of mathematics. The permanence and universality of mathematics derive from its abstractness; its apparent necessity, or "fatedness," is a concomitant of the rigidity of formal logic.

Neither the necessity nor the universality is taken for more than a temporary appearance by those who believe mathematics and logic to be of purely human origin. Others, including many who believe that numbers were discovered rather than invented, find in mathematics irrefutable proof of the existence of a supreme and eternal intelligence pervading the universe. The former regard mathematics as variable and subject to change without warning; the latter see mathematics as a revelation of permanence throughout eternity, marred only by such imperfections as are contributed by the inadequacies of human understanding. With constant progress toward clearer perception of the infinite, all blemishes will disappear, and mathematics will shine forth as a flawless embodiment of the eternal truth.

The first hint that such a creed might be rationally possible appeared in the sixth century B.C. in about half a dozen simple statements concerning straight lines and circles, some of which Thales is said to have proved.

If a straight line be drawn through the center of a circle, the circle is divided into two parts identical in all respects. Again, if two sides of a triangle are equal, the angles opposite the equal sides are equal. These two propositions are obvious on drawing the corresponding figures, and likewise for this: if two straight lines intersect, opposite angles at the point of intersection are equal in pairs. By merely using our eyes we see that these geometrical facts are "true." And if we reflect a little, we "see" with our minds that these facts do not derive their truth from any particular figures we may have drawn, but are conceivably true of any circle, any isosceles triangles, and any pair of intersecting straight lines humanly imaginable. That is, within their own realm, these "truths" are universal. Why? Some say it is all merely a matter of definition. Others find solace in the belief that the "universality" of abstract lines is an attribute of the Eternal Mind.

A fourth proposition is equally obvious: if a rectangle is inscribed in a circle, each diagonal of the rectangle passes through the center of the circle. This, it may be granted, is not very inspiring. But restated in the following equivalent form, it becomes what many have considered the most beautiful theorem in elementary geometry: the angle inscribed in a semicircle is a right angle. The invariance, the constancy of the angle, no matter at what point of the circumference of the semicircle its vertex may be, excited the wonder of Dante.

Each of these four propositions is intuitively evident on inspection of a simple figure such as might be scratched in the dust by a child at play. All four seem to have been known long before the sixth century B.C., when, for the first time in history, they were really looked at, not with the unwondering eyes of a child, but with the questioning mind of a reasoning man.

Like all the thousands of unthinking spectators who had seen the obviousness of these propositions, Thales also saw that they are intuitively evident in the sense of visual "truths." Then, possibly, he began to question the obviousness, the necessity, of these simple geometrical facts. What does it mean to say that a statement about a figure composed of lines is true? If Thales did not actually put the question to himself in this form, or explicitly in any form, he proceeded to act as if he had. For what he did we have to rely on the statements of Greek historians who wrote long after Thales had ceased troubling about straight lines and circles. The histories are concise to the point of ambiguity; but their import is that Thales introduced abstraction and proof into the study of lines, both straight and curved. Proof gave meaning to truth as it appears in geometry. It remained for Plato and his school to imagine what might give meaning to proof.

Geometry in Egypt and Babylonia had not outgrown its origin in the immediately practical when Thales transported it to Greece. It was still chiefly concerned with empirical rules for the computation of areas and volumes. Such a proposition as the one about the pairs of equal angles made by intersecting straight lines would scarcely have occurred to the practical minds concerned with pyramid building and canal digging. Yet this proposition is frequently required in proving others which are neither obvious nor useless. The like holds for the idealized, abstract lines of the geometers, which certainly would not appeal to a naively practical mind as worthy of serious thought. In passing from the concreteness of sensory experience to the abstractness of ideal constructions, Thales took a stride into the infinite, leaving his contemporaries thousands of years and a universe behind him.

His second advance, equally epoch-making, was to imagine that some of the abstractions of geometrical facts of common observation might be deducible from abstractions of simpler facts of the same general kind. He is said to have "proved" some of his theorems in the "perceptible," "intuitive," or "sensory" manner of the Egyptians, that is, roughly, "by eye." Others, and this is the distinction of cardinal importance for the development of science, mathematics, and philosophy, he is credited with having "proved," or having attempted to prove, in the "abstract," "general," or "universal" manner of the classical Greek mathematicians. A liberal interpretation of the last is justified by the circumstances under which it was written. It was addressed to Greek mathematicians long after the time of Thales. To these men the Greek manner of proof could mean only strict deductive reasoning.

Not to award Thales more honor than he might justly claim, it should be mentioned that some historians admit that he correctly used deductive reasoning but did not recognize the generality of the process—the explicit statement of assumptions, followed by the strict deduction of consequences. Without more evidence than has yet been adduced such a conclusion cannot be refuted. Neither can it be sustained. A part of the credit for the invention of proof in mathematics is not denied Thales by any competent critic. Full honor for having developed the deductive method is awarded to the father of western number magic, Pythagoras. Thales seems to have had no magic in him, only reason and a most wonderful common sense. Two further items from the irrecoverable past may close his account with both mathematics and philosophy.

The interest of the first item at this point is mainly historical. Its significance for the early development of Greek philosophy and mathematics will appear in connection with Zeno and his exasperating paradoxes.

The Great Pyramid was the wonder of wonders of the ancient world. Like every Greek tourist to Egypt, Thales dutifully viewed this most impressive monument to a king's mummy. The degenerated priests exhibited it to their visitor as their final demonstration that Egypt had forgotten more civilization than Greece would ever know. If the but recently civilized Greek was awed by the colossal bulk that threw its pointed shadow far out over the sands, he managed to dissemble his astonishment. To the confusion of his hosts, Thales casually proceeded to measure the height of their stupendous pyramid. Dazed by his profane audacity, the priests were unable to imagine how a mere mortal could so effortlessly accomplish the seemingly impossible. For hundreds of years the heads of all their kind had been as void as the mummified king's empty cranium of any brains that really mattered. Centuries of mumbling over the Book of the Dead had atrophied their awareness of the once-living; and the past greatness of their own builders was no longer within their memory. Egypt was through; Greece was about to begin.

There are two principal versions of how Thales performed the miracle. The simpler states that Thales measured the shadow of the pyramid when his own shadow was equal in length to his own height. Anyone who remembers a little school geometry will see how this gave him the solution of his problem.

The second version is almost the same. It leads directly to the general statement of one of the most useful theorems in all geometry. To recall it, the triangles ABC, PQR are such that the interior angles at A, B, C are equal respectively to the interior angles at P, Q, R. The length of the side joining A, B is written AB. Thus AB denotes a certain "number" expressing the length of the side; likewise for the other sides of both triangles. The theorem asserts that the fractions

$$\frac{AB}{PQ}, \frac{BC}{QR}, \frac{CA}{RP}$$

are equal.

Thales is credited with a proof of this theorem. He might indeed have proved it for triangles whose sides are measurable by common whole numbers. He could not possibly have proved it when such is not the case, as the "numbers" then required to measure the sides were not imagined until after he was dead.

The uncovering by Pythagoras of a single specimen of the appropriate numbers was a major turning point no less in the development of mathematics than in the evolution of metaphysics. Its influence on both will be noted in the proper place; for the moment we observe a feature of Thales' supposed proof that is of more than passing interest. It is this: a proof that convinces the greatest mathematician of one generation may be glaringly fallacious or incomplete to a schoolboy of a later generation. A properly taught pupil in high school today can detect the flaw in any proof such as Thales might have given. He can do so, however, not because he is a keener mathematician than Thales, but because some of the ablest mathematicians in history, in the three centuries immediately after Thales, followed the way of abstraction and deduction pointed out by him.

The second item of interest discharges an earlier obligation. It was remarked in connection with deductive proof that the ancient Egyptians may have had some idea of it, and that the doubtful evidence for this possible anticipation rests on the testimony of one man. The man in question is Democritus (c. 460-c. 362 B.C.), the aggressive protagonist of atomism. Nicknamed "the Laughing Philosopher," Democritus started out in life a very rich man, saw the world in grand style, got rid of his wealth to the last penny, and laughed himself to death at the age of nearly a hundred, some say actually on his hundredth birthday. As the following thumbnail autobiography hints, the Laughing Philosopher was not distinguished for his modesty.

"I have wandered over more of the earth than any other

man of my time," he begins, modestly enough, "investigating the most recondite matters. I have studied many climes and many lands, and I have listened to numerous learned men. But never yet," he continues, warming to himself, "has any man surpassed me in the construction, *with demonstration,* of lines—not even the Egyptian rope-stretchers, with whom I sojourned all of five years."

As for "with demonstration," Democritus may have known more about Egyptian mathematics than has yet been unearthed; or he may just have been enjoying a sardonic laugh at the expense of his revered countryman Thales. For if the Egyptians actually proved anything in their empirical geometry, it is more likely that Thales heard of proof from them than that they learned what he had invented and began practicing it almost immediately. They were not, so far as is known, celebrated in the sixth century B.C. for their love of abstractions; and it seems unlikely that they would have seized on any that might come their way. If Democritus was not perpetrating a sarcastic hoax, he may have been suffering from a confusion of memory in his old age. Whatever may be the fact, Thales will continue to stand immovably at the beginning of western thought in both mathematics and philosophy—and this despite his assertion that "Everything is full of gods."

CHAPTER - - - - - - - - - - - - - - 8

One or Many?

THALES was followed by Anaximander (610-546 B.C.), who may have been a personal disciple. An important link in the long chain of mathematical philosophers from the sixth century B.C. to the present, Anaximander has a special claim on our attention in his conception of the infinite. It would be reading the present into the past to discern a hint of the mathematical infinite in his "apeiron." Yet this elusive concept as explained by ancient commentators had some of the qualities that we now ascribe to the infinite. It definitely started western speculation regarding a possible limitless, eternal universe.

It may be noted in passing that the unbounded is not necessarily infinite. The surface of a sphere, for instance, is of finite extent though unbounded. And in one of the space-time models of the physical universe suggested by the theory of relativity the universe is unbounded and finite.

Few details of Anaximander's life have survived. Cicero makes him a friend and companion of Thales. In any event he was well versed in the Thalesian geometry and philosophy and passed them on, with additions of his own, to Pythagoras. He was one of the earliest if not the first of the Greeks to commit his science and mathematics to writing. His work is known to us only through fragmentary allusions by ancient historians and philosophers who, as usual, frequently contradict one another. Anaximander also has the unique distinction of being the first man on record to deliver public lectures on philosophy. At one time he probably conducted a school for boys. The only story that has lasted pictures him as a conscientious pedagogue. "For the sake of the boys I must try to recite better," he declared when he was jeered at for his custom of intoning his teachings.

The ambition of Anaximander's admirers was to have him beat his master's record in everything. Thales had foretold an eclipse; Anaximander predicted an earthquake—a feat the seismologists of the twentieth century have yet to duplicate. Thales said, "Everything is water." Anaximander went him one better; everything for him was water and mud. Incidentally, Anaximander thus inaugurated the profitable tradition of the pupil in philosophy contradicting his master. Thales had not ventured to explain how the world came to be as it is; Anaximander gave science its first comprehensive theory of evolution. Naturally it was not a very plausible theory, but it took a step in the direction of naturalism and away from supernaturalism in the explanation of nature. We shall not pause over any of its details, as Empedocles produced a more amusing account of the origin of living things which will be noted later.

Among his numerous "firsts" Anaximander made the earliest map of the world as known in his time. Egyptian and Babylonian maps of restricted localities may have inspired Anaximander's attempt to mark off the land from the water on an all-inclusive scale. Where his predecessors had seen the special problem, he imagined the general. If only Pythagoras had not gone numerological, Greek science might have developed from such promising origins more rapidly than it did, and quite possibly would have accomplished much of lasting value. Another of Anaximander's "firsts" was his reputed formal exposition of geometry. It cannot have been extensive; but if, as has been claimed by some, its few theorems were arranged in a logical sequence, it was an epochal work.

In astronomy he used the gnomon (carpenter's square) in determining the meridian and the places of the solstices. He also fathered a theory of the heavenly bodies which in modified form passed through the cosmology of the Pythagoreans, thence into Plato's and those of his debased successors, to come at last to eternal rest, in one of its details, in the unworkable theory of celestial vortices proposed by Descartes (1596-1650) to account for the motions of the planets. The heavenly bodies, according to Anaximander, are globes of fire and air, and each carries with it a living fragment of the deity. In some degree it is therefore a god. Descartes' planets, while not themselves gods, were set in motion by the deity, who at the creation imparted motion to all matter. In the rhythmic revolutions of Anaximander's planetary spirits we almost catch the all but inaudible notes of that "music of the spheres" which first becomes clear and harmonious in the astronomy of Pythagoras.

Anaximander also fixed the Earth where it was to remain for over two thousand years, till (1543) Copernicus (1473-1543) displaced it from the center of the universe. This, however, was probably not entirely his own accomplishment. With unprecedented daring Anaximander undertook to estimate the size of the Sun. Though his data and his instruments were inadequate for the task, and his conclusion was badly mistaken, he deserves full scientific credit for his direct appeal to nature. It may be true, as a great scientist of the seventeenth century observed, that "the book of nature is writ in mathematical symbols," but it requires more than a knowledge of mathematics to read that book understandingly.

Anaximander's "Infinite" was the subject of inconclusive controversies even in ancient times. Plutarch, agreeing with Aristotle, said it was simply matter. Anaximander himself is reputed to have described his Infinite as permanent in its whole though variable in its parts, the inexhaustible source of all things, and the eternal all to which they return. This led to an irresistibly seductive speculation: in the ceaseless unfoldings of the Infinite, evolution may have run its course many times, perhaps even infinitely often, leaving "not a wrack behind" of all the perished worlds and extinct races time has known. Possibly this poetic jest of old Anaximander's imagination inspired the Pythagoreans, and after them Plato, to their dream of the Eternal Recurrence. Another possible source of this infinite nightmare will be noted in connection with Pythagoras.

More than a hint of two of the most protracted debates in all metaphysics emerges from Anaximander's all-generating, all-absorbing Infinite. Is the universe a One or is it a Many; is it a "being" or is it a "becoming"? From Pythagoras to Parmenides, from Parmenides to Zeno, from Zeno to Socrates and Plato, and from them to a host of mystics, logicians, metaphysicians, theologians, and mathematicians down to the twentieth century, these interminable disputes in one guise or another have occasioned an appalling mass of controversy. Not all of it by any means was barren of positive achievement, especially in the mathematical theory of the infinite.

Although we shall never know what kindled a mind like Anaximander's to glow with some perception of problems which were to engage his successors for centuries after his death, we may perhaps find traces of his least unreasonable imaginings in the prescientific myths and fables which his feeble science struggled to rationalize. The characters in Plato's dialogues, we recall, when hard pressed to substantiate some wildly unscientific fable concerning matters of observable fact or imaginative inference, frequently cited unnamed wise and good men and women of an indefinite past as witnesses to the truth. These ultimate authorities appear to have been fabulous only in that they were fables. They actually existed as the mythologies which Thales and Anaximander had striven to supplant by less incredible inventions of the human imagination. Rather than accept the responsibility for these partly discredited myths sophisticated in their own attempt at science, the cautious philosophers referred them to a sacred past, and in so doing invested them with the customary veneration accorded the ancient dead. Distance in time conferred the distinction which in our own day is usually reserved for mysterious sages from the antipodes.

In Anaximander's century an explosive outburst of Orphism —the tangled mass of myths, rationalisms, and religious doctrines originating, at least in its Hellenized form, in the tragic story of Orpheus and Eurydice—obscured any attempt at a scientific account of nature. That such an incoherent mythology could have influenced the course of scientific speculation for centuries may seem incredible to us now; but that it did appears to be the fact. The first crude theories of evolution, those of Anaximander and Empedocles, were obvious attempts to rationalize the Orphic myths of creation with their dismembered and reassembled gods. On another level, the Pythagorean belief in the transmigration of souls and the depressing creed that this life is a punishment for the sins of some previous existence and a possible purification for life in a better world to come are unadulterated Orphism. It might almost be said that if we strip an ancient science of its rational trappings we invariably find a more ancient myth. What is perhaps more significant is that some ancient myths, also some not so ancient, when observed in the nude are trivialities of common arithmetic or elementary geometry. This will appear when we come to Plato.

CHAPTER - - - - - - - - - - - - - - 9

A Dream and a Doubt

IF ONE man more than another is to be credited with starting the mathematical and physical sciences on their course from antiquity to the present it is Pythagoras. And if "western civilization" means the technology and commerce of recurrent industrial revolutions detonated by the application of experiment and mathematics to the physical world, Pythagoras was its prime mover. All this is on the strictly scientific side. On the side of purely intellectual activity, the numerology (number mysticism) of Pythagoras and his Brotherhood is the source of essential germinal ideas in Plato's metaphysics of the sciences.

Standing in the sixth century B.C. at the decisive rupture between oriental mythologies and occidental rationalisms, Pythagoras looked both before and after. Behind him, as far as he could see, a rational mentality kept struggling to emerge from a stifling past of age-old superstitions, crude magic, and unrestrained number mysticism. Before him, he imagined a future of reasoned enlightenment, experimental science, and mathematics. Incredibly old and already dying in some of its members, the mythological past was slowly sinking out of memory. Barely imagined from a few significant hints in his own work, the future which Pythagoras visioned may have seemed brighter with promise than actually it was to be. Which should possess his mind, the mystical past or the rational future? As may have been inevitable in his day, the decision was a fatal compromise.

Neither all mystic nor all rationalist, Pythagoras was a singular fusion of the two, combining in one personality the credulous eagerness of a child for the miraculous and the mysterious with the patient humility of the mature scientist who is willing to learn by experience and to abide by its teachings. As an experimentalist he glimpsed the power and the utility of numbers in the description of natural phenomena. As a mystic philosopher he extrapolated his scientific success to the astounding generalization that everything is number, possibly the most mischievous misreading of nature in the history of human error. Occidental science and occidental numerology, as ill-assorted a pair of twins as ever were born, thus sprang from one source, the mind of Pythagoras.

Both science and numerology continue to thrive after twenty-five centuries of fighting each other, and neither as yet shows any evidence of being strong enough to destroy its hated rival. If numerical superiority counts for anything, active or potential believers in numerology outnumber believers in science thousands to one. In western civilization the numerology is not necessarily of the sorry fortune-telling variety, though in even the most advanced civilizations this prostituted arithmetic is common enough. It assures anyone who will adjust his conduct to his true Pythagorean number health and prosperity in this life, to be followed by everlasting joy and felicity in the next. But in general, modern Pythagorean numerology is much more refined, and it may be an unintended slur to call it numerology at all. The more subtle manifestations of the ancient doctrine are disguised and occluded in those monumental philosophies that incorporated fragments of the Pythagorean "everything" into their foundations.

In science also the creed that everything is number has been successively refined to accommodate advancing sophistication. Today no reputable scientist would risk asserting that everything is number lest his colleagues think him queer. If he did have a secret hankering to restore number to its Pythagorean universality, he would not phrase his declaration of servitude to the past so bluntly. It would suffice—as it already has—to go back no farther than Plato, who is said to have asserted that "the deity ever geometrizes." Without jeopardizing their scientific reputations, the modern Pythagoreans might profitably announce—as indeed Sir James Jeans did in 1930—that "the Great Architect of the Universe now begins to appear as a pure mathematician." This is a step ahead of "everything is number," but only a step; for the mathematics applied to the architecture of the universe is based on numbers. Pythagoras would have understood this modernized version of his creed. He might even have certified the sublime truth it may express.

The most enduring residue of Pythagorean numerology is only remotely connected with arithmetic. Briefly, it is the very human desire to find easy shortcuts to positive knowledge. Laborious experiment to discover the facts about our environment is wearisome to all but a persistent few. Surely all this blundersome experimenting can be by-passed by some more direct route to the heart of nature? Certainly it can, say the numerologists of today, as their predecessors have insisted for the past twenty-five centuries.

Numerology is the faith that the universe can be summed up and compressed through a single grand formula to a unified whole comprehensible by human beings. Thorough understanding of the one supreme generalization will make all the secrets of nature plain. The tyranny of time will then be overthrown, and man will become the undisputed master of his future.

Such is the dream. With every advance of verifiable knowledge it fades farther into the unknown. Discovery seems to nullify itself in ever vaster horizons to be explored. Pythagoras believed he had found the tremendous formula in his "everything is number." As more of the physical universe was revealed by controlled observation, "everything" was successively whittled down to less immoderate proportions. By the twentieth century "everything" for Sir Arthur Eddington and his school had shrunk to mean only all the laws of the astronomical and physical sciences. But in its earlier forms the vision of ultimate knowledge included literally everything, from the heavens to the human emotions. When Pythagoras announced that everything is number he meant exactly that.

Probably no scientist today hopes to include this universal everything under the rubric of number. Others, the orthodox and immovable adherents of the ancient wisdom, do not need to hope. They are as certain as ever Pythagoras himself was that everything is number.

It will be well here to mention ahead of its historical place a sinister doubt that is said to have troubled Pythagoras in his last dream. The same doubt has returned in the twentieth century to perturb the modern Pythagoreans.

In the usual sense of conformity to reason, the Pythagorean philosophy, including its numerology, is strictly rational. From certain assumptions conclusions are deduced by a cold and relentless logic that compels assent. Once the assumptions are accepted, it is futile to rail at the alleged absurdity of the conclusions.

The basic assumption underlying all applications of number to science is that the laws of nature are rational. That is, these supposed laws are assumed to be accessible to a sane mind, and to be expressible in terms comprehensible by human reason. They may not be.

This is the grim possibility that caused Pythagoras to question his own sanity and to doubt his grand formula for solving the universe. Happily for what was to be the future of science, doubt visited him at the end of his career rather than at the beginning.

As it has been stated, the doubt may seem to cancel itself. If the "laws of nature" are forever inaccessible to the human reason, they cannot be of much importance for human beings, whatever they may be. "The Unknowable" that Herbert Spencer talked about so knowingly may safely be ignored. But there is a sense in which the doubt seems to mean something: all the "laws" which we have imagined were natural necessities may have been put into nature by ourselves. Instead of taking we may have been giving.

Whether the Pythagorean dream of complete and final knowledge of the universe is sublime or not, or whether following it has benefited our race, is not for us to judge. We are concerned primarily with what induced the dream, and incidentally with some of the men who in their zealous pursuit of it found things whose value for millions of human beings has endured. By almost any standard Pythagoras was the first of these and one of the greatest. Before passing to his work, we shall report what kind of man he seemed to be to his contemporaries and followers in antiquity and what kind of life he lived.

CHAPTER - - - - - - - - - - - - - - 10

Half Man, Half Myth

AS BEFITS a sage who knew that he was great and imagined he was half divine, the legendary Pythagoras is an austere figure, always wise, always temperate, and never once relaxing into anything half so human as the engaging rascalities of Thales. When asked by his disciples what they should call him, Pythagoras did not arrogate to himself the title of wise man, but requested that he be called simply a philosopher—a lover of wisdom. This sincere modesty is as exact an index as anything, in the devious life of this philosopher, to his almost fanatical devotion to knowledge and wisdom. His was a genuine humility in the presence of the knowable.

It may be stated once for all that this extraordinary spirit is known to us only through legends and traditions, for none of which is there any contemporary documentary evidence. Even his dates are in dispute, but 569-500 B.C., accepted by many scholars, are frequently given. Slight adjustments of both dates seem necessary to fit the chronology of his life, and these are usually assumed by the chroniclers without comment.

Though it is unlikely that anything fully reliable will ever be discovered about Pythagoras as a man, a great deal is known of what his successors thought of him. As in the case of Thales, these unsubstantiated judgments may be more revealing than any official biography. Invariably they represent Pythagoras as an outstanding figure even among the great. The common man, so-called, may know almost nothing of science. Yet he seems instinctively to recognize a scientist of absolutely the first rank when, at intervals of centuries, such a one appears. As a memorable instance even the most highly cultured of Newton's contemporaries were incapable of understanding his epochal achievement. Yet somehow they, and the man with no pretensions to culture, knew that here was a scientific mind without a superior in history. And when Einstein appeared, the same instinct for a revolutionary advance in science again awoke, though only one in thousands could follow the mathematical technicalities of relativity. Mere talent and what in an average century is first-rateness never evoke this instinctive response.

Pedants, intellectual snobs, and worshippers of the second-rate may decry this popular recognition of the highest achievement as but one more proof that the public loves a sensation. But for all their envy, they cannot repress the sound instinct of their wiser fellows for supreme greatness. And for his part, the man who knows little or nothing of science offers the master his tribute of an anecdote, more likely than not without foundation in fact, which epitomizes what the great man has meant to him.

So it was with Pythagoras. Universally recognized as a master among masters, he lived far beyond the narrow confines of his own body, in the wonder and respect of his unlearned fellow men. The legends of his life are not about

Pythagoras; they are Pythagoras, and it matters nothing at all whether every last one of them is false or whether it is true. From all the hundreds of doings and sayings attributed to Pythagoras, anyone may accept or reject what he pleases. Those legends that accord with a particular individual's conception of the man's essential greatness will be acceptable to that individual. The others may be rejected as the stupid fabrications of dullards incapable of appreciating the master.

Even in antiquity Pythagoras was a dim and legendary figure. Aristotle, for example, born in 384 and dying in 322 B.C., was only about two centuries later than Pythagoras; yet he seems to be in some doubt whether such a human being as Pythagoras ever existed, mentioning him by name only twice. Rather than commit himself by referring to the teachings of Pythagoras, Aristotle cautiously attributes the sound mathematics, the music, the harmonious astronomy, and the fantastic numerology, traditionally ascribed to the master himself, to anonymous Pythagoreans. The very name Pythagoras, recalling the Greek for one who is inspired, seems to have been regarded by the over-suspicious as a feeble pun on the Greek (python) for a soothsayer. It followed for the rigidly skeptical that Pythagoras was not a man but a nameless oracle.

Aristotle was justified in being moderately cautious in crediting Pythagoras himself with specific discoveries. For it is almost certain that many of the things branded with the master's name were the inventions of his disciples. Some indeed were made long after Pythagoras, either as a man of flesh and blood or as a nebulous hypothesis, had passed on to a stage of existence other than the human. Even during his supposed mortal lifetime Pythagoras was given the credit for all discoveries made by his disciples, somewhat as the director of a scientific research laboratory today occasionally monopolizes the publicity for his entire staff. But however remote Pythagoras had become by the time biographies of him appeared, the massed testimony of ancient historians establishes his material existence for the majority of critics beyond any reasonable doubt.

The unique contemporary notice of Pythagoras that has survived is a sour tribute by the misanthropic philosopher Heraclitus. This renowned lover of wisdom flourished about 500 B.C. He was nicknamed the Weeping Philosopher, and, appropriately enough, is remembered in histories of philosophy for his too hasty generalization that "All things flow." Evidently suffering from a twinge of professional jealousy, Heraclitus has this to say of his more successful competitor for immortality: "Pythagoras, son of Mnesarchus, has pursued research and inquiry more assiduously than any other man. He has compounded his wisdom from polymathy and bad arts."

This at any rate preserves the name of Pythagoras' earthly father. Mnesarchus was a stonecutter of Samos, where Pythagoras was born at some doubtful date between 580 and 569 B.C. Beyond a vague rumor that Pythagoras' mother was of Phoenician extraction, little is known of her except that she is said to have accompanied her wandering son on his last journey.

Like some others of the major prophets of our race, Pythagoras to his disciples was divine. His heavenly father was generally believed to have been Apollo. In proof of his celestial descent, Pythagoras when properly approached would display a golden thigh. This curious legend is so persistent that it may be the miracle-mongered distortion of some real physical infirmity.

The acid tribute of Heraclitus merits serious consideration as the unsolicited testimonial from one lover of wisdom to another. "Polymathy" means simply encyclopaedic knowledge, surely no serious disqualification for a man who undertook to sum up the universe in a single formula. But it is obvious that Heraclitus was paying no compliments. By "polymathy" he seems to have intended an unpraiseworthy eclecticism, the implication being that Pythagoras lifted whatever he fancied wherever he found it and succeeded none too well in assimilating his thefts.

If Heraclitus was right and not just envious, Pythagoras was merely an inferior and rather disreputable magpie. But Heraclitus had the same opportunities as Pythagoras to make something of all the knowledge and wisdom lying about everywhere in that opulent sixth century, and he made little or nothing of what he might have had for the taking. Pythagoras may have appropriated everything he could seize, but he did not stop at that. The rough stones he gathered were gems when he passed them on. And to close the account between these two contestants for a place in human remembrance, it is the fact that for anything of significance contributed by Heraclitus to life as now lived, Pythagoras outranks the disgruntled philosopher in approximately the ratio of infinity to one.

All legends of Pythagoras make him a tireless traveler until he was well into his middle years. It is not recorded at what age he left his native Samos; but it is said that as a youth of eighteen he fell in with Thales. If he did not actually become one of the Wise Man's pupils, Pythagoras absorbed the Thalesian philosophy and mathematics at second hand from Anaximander. He must have been properly impressed; for when Anaximander assured him that the true wisdom was to be mastered only in Memphis, Pythagoras immediately set out for Egypt without a penny in his wallet. A more romantic but less credible legend pictures Thales himself initiating Pythagoras into the mysteries of Zeus on sacred Mount Ida and urging the young man to get to Egypt as fast as he could, even if he had to walk the whole way.

Some accounts credit Pythagoras with all of twenty-two years among the learned men of Egypt and Babylon. Others have him wandering restlessly all over Egypt, Mesopotamia, Phoenicia, India, and even Gaul, far beyond the Pillars of Hercules, and declare moreover that he absorbed all the knowledge and wisdom of the Hebrews, the Persians, the Arabs, and the blue Druids of Britain.

From what is now known of pre-Greek mathematics in Egypt and Babylonia it is not improbable that Pythagoras learned much about numbers and figures from the slowly expiring civilizations of the Near East, regardless of whether he ever lived among their peoples. The number magic that he brought back with him to Samos is almost as valid evidence of travels in the East as a canceled passport. And though he may never have set foot in India, his missionary zeal for the doctrines of reincarnation and the transmigration of souls is enough to prove that he must have studied under some master thoroughly versed in the religions of the Orient. It does not seem to be definitely known how far west these doctrines had penetrated in the sixth century B.C., nor, for that matter, where they originated. One thing seems certain: they were foreign to the Hellenic genius when Pythagoras incorporated them into his own teachings. In a somewhat cruel form they passed thence into the eschatology of Plato, who added the characteristic touch that the souls of cowards are reincarnated in the bodies of women. The souls of the stupid, according to Plato, animate four-footed beasts and birds; while an utterly worthless soul, being unworthy to breathe pure air, must content itself with the body of a fish. Pythagoras was more merciful.

It is interesting in reviewing the life of another, if not one's own, to observe the critical points where a slight deviation from the path actually followed might have led to complete success instead of to partial failure. Pythagoras passed through one such point when he decided to leave Samos and study in Egypt. Had he not left his fellow Greeks to their accustomed ways while he sought knowledge and wisdom in the East, his name might be as unknown to us as those of the thousands who stayed at home and lived out their simple lives in ignorance and peace. In all there were three decisive turning points in this fated man's life. The second changed his course directly toward the final catastrophe. At the age of forty (about 530 B.C.), Pythagoras returned to Samos.

His fatal blunder was that which has undone many a prophet. He attempted to raise his own people to his own level. At the height of his enthusiasm for all the mysticism, the mathematics, and the number lore he had acquired, Pythagoras immediately set about the enlightenment of those he had left behind to perpetuate the bucolic traditions of their rustic forefathers. In his pathetic simplicity he engaged the municipal amphitheater. Expecting to see it crammed to the sky, the sanguine philosopher found himself staring into a gulf of empty stone benches.

The populace was more than merely indifferent. It was thoroughly roused and enraged. What could Pythagoras, the son of old Mnesarchus the stonecutter, know about anything? Why, men of his own age even remembered him as a little boy running about his father's workshop, getting in everybody's way, and pestering the customers with stupid questions. He was always asking about the silliest things of no practical use to stonecutters or anyone else. Was it believable that this obnoxiously ignorant little boy could have grown up into a mature man with sound sense in his head? It was not. As an idle adolescent always loafing about the street corners and waylaying the leading citizens for interminable cross-examinations on matters of no importance, this fellow who now called himself a lover of wisdom was even more insufferable than he was as a boy. Let him love wisdom as hotly as he liked; they had their work to do and needed a good night's sleep after the long day's toil and sweat. It was just like his conceit to put his proposed lecture after supper when everybody was tired and full of meat.

Then too there was the ridiculous incident of the chastised dog. A man had a right to beat his dog. It was his, and he could do what he liked with his own property for any good reason or for no reason at all. But this crazy crank Pythagoras had made quite a fuss when he saw one of the prominent citizens taking a stick to his dog. "Stop beating that dog!" he had shouted like a madman. "In his howls of pain I recognize the voice of a friend who died in Memphis twelve years ago. For a sin such as you are committing he is now the dog of a harsh master. By the next turn of the Wheel of

Birth, he may be the master and you the dog. May he be more merciful to you than you are to him. Only thus can he escape the Wheel. In the name of Apollo my father, stop, or I shall be compelled to lay on you the tenfold curse of the tetractys."

So his father was now Apollo, was he? Since when had that doddering old Mnesarchus, ready to stumble into his grave at any moment now, been one of the immortal gods? This impostor Pythagoras was worse than a nuisance; he was as mad as a goat with a broken head. What right had he to go about scaring healthy people into their beds with his foreign curses? If the owner of the dog died, they would know what to do with the man who had killed him. In the meantime he could lecture to the wind.

There is no record of what Pythagoras said—or thought—of his reception by his own people. Unlike another famous teacher he did not vent his disappointment in petulance. If they would not come to hear him, he would take his message to them. He left the empty amphitheater and got himself a pupil—just one, and poverty-stricken at that. Under the circumstances he might have been pardoned had he said, "Let him which is filthy be filthy still!" But Pythagoras really was a philosopher, and he knew that one of the ways of loving wisdom is to share it with others. What he yearned most ardently to share was his passion for geometry as a deductive science.

Going far beyond Thales, Pythagoras had discovered and proved numerous theorems in what is now the first course in school geometry. Always remembering that some of the theorems attributed to Pythagoras may have been the discoveries of his disciples, we can assert, on the authority of

Greek historians of mathematics, that Pythagoras cast geometry in the shape it was to retain for about two thousand years. He is credited with having recognized that definitions must be set out at the beginning of the entire subject, and that the postulates (axioms) from which deduction is to proceed must be stated explicitly. Further, he strove in his own proofs to guard against the surreptitious entry of further assumptions in addition to those of the admitted postulates.

It was like a game: here are the pieces; only certain rigidly prescribed rules for moving them are permitted; what are the possible configurations of the pieces in an honestly played game? The pieces are the definitions and postulates; the rules for the moves are those of formal logic; the possible configurations are the deductions from the postulates by means of the logic: and the outcome is the theorems of geometry.

When fully formalized any mathematical argument proceeds by the same tactic: definitions and postulates; deduction; theorems. The rigidity of the Greek technique, but not its underlying logical justification as one deductive science among many, was relaxed in the creation (1637) by Descartes of analytic geometry, in which all the machinery of algebra and mathematical analysis is applied to geometry. The gain in power and simplicity was tremendous, and the strict Greek model passed out of use. But the underlying philosophy remained.

To make it worth his impecunious pupil's time to play the game, Pythagoras paid him a penny for each geometrical theorem mastered. This suited the poor young man perfectly. By merely sitting in the shade, using his eyes and listening attentively he earned better wages in an hour than he could have made in a full day of back-breaking labor in the broiling sun. But Pythagoras, disciple of the tight-fisted Thales, was not letting the pennies get away from him never to return. Just as his pile began to grow to a respectable sum the pupil, in spite of himself, became avidly interested in geometry and begged his teacher to go faster. The irrepressible Greek in Pythagoras saw his chance and reached for the stakes. Confiding to his pupil that he was a desperately poor man himself, Pythagoras suggested that the pupil now pay him a penny apiece for each new theorem. By the time the young man had absorbed all the geometry he could hold and was ready to return to hard work, Pythagoras had gained back all his money and had exactly as much geometry left as when the game began.

It must be admitted that the last of this story is hardly in character with the masters' habitual austere integrity. It may be a late fable invented to dramatize the fact that it is impossible to diminish intangibles by subtracting intangibles from them, or by sharing them with others.

Assured that his pupil was now thoroughly indoctrinated with the new truth, Pythagoras made a second and last effort to enlighten his own people. Always as shrewd in psychology as he was wise in geometry, the master revised his grand strategy. His failure had been his own fault. He should have returned to his native town not as a professor of knowledge and wisdom, but as a pompous mystagogue endorsed by one or more of the leading oracles. Announcing that he was leaving Samos to master the sacred mysteries in Delos and Crete, Pythagoras promised his detractors that he would return as soon as he had acquired the proper credentials for instructing them in matters of the highest practical value.

He kept his word. Possibly on his visit to Crete he learned

something himself. For there he may have heard of Epimenides the Cretan. Epimenides the Cretan is justly immortal for his cynical remark, "All Cretans are liars." Was he lying when he said that? Or was he speaking the truth? Either answer contradicts itself. This was the first of several logical paradoxes that were to perplex the Greek philosophers and mathematicians. If Pythagoras heard that Epimenides had made that disturbing statement he might well have felt uneasy for the security of some of his own. It seemed to slip through the mesh of deductive reasoning like an eel. Was such reasoning as powerful and as sound as the master had thought?

Perhaps it was fortunate for the futures of science and mathematics that Pythagoras either did not hear what Epimenides said or ignored it. Otherwise all the subtle logical difficulties, the "Epimenides" paradox among them, that appeared in the foundations of mathematics toward the close of the nineteenth century A.D., might have prevented Pythagoras from laying the cornerstone of all mathematics in the sixth century B.C.

On his return from ostensibly consulting the oracles, Pythagoras found the Samians a little less unfriendly. After all they were only human. In spite of their hostility toward their would-be uplifter they began to become curious about him. There was a rumor that he had subdued an enormous bear, that was ravaging the communal pigsties, by simply pointing his finger at the beast and commanding it in the name of Apollo to desist. Then, too, there was all this talk about what he ate, or rather about what he would not eat. What could the man possibly have against beans? They were a staple of everyone's diet; and here was Pythagoras refusing to touch them because they might harbor the souls of his dead friends. Did anyone ever hear of such nonsense? He had even deterred a cow from trampling a patch of beans by whispering some magic word in its ear. Absurd!

But his ban against eating the flesh of animals might be worth looking into. For who could say that the souls of the dead did not pass into the bodies of animals when there was a shortage of new human bodies to accommodate all the souls suddenly released in a battle? Pythagoras himself, though not claiming openly to have inhabited the bodies of animals in previous incarnations, insinuated that even he might have done so for his impieties. His descriptions of some of the lives he had lived in human or divine shape were singularly detailed and eerily convincing. Horrible dreams of their own, when remembered and analyzed in the unearthly light of this sudden new knowledge, hinted that the transmigration of souls might be the dreadful fact Pythagoras said it was. What if it were? The thought of all the souls they might have left shivering in the void by devouring their own goats and swine made the good Samians extremely unhappy. A few weeks more of these upsetting suggestions, and they would all be strict vegetarians—except for beans.

Equally upsetting was the ghastly thought that some of their own children might be malicious little monsters with no souls to restrain their bestial instincts. For Pythagoras had assured them that the total number of souls in the universe is constant. Perhaps he was right in scolding them for having such large families and warning them never, no matter how compelling the urge, to have more than ten children apiece. There was something about the number ten that made an eleventh child repeat all the disagreeable characteristics of the first. They did not understand this very clearly, but it must be true—"Himself said it." For all of a week they conducted their lives as Pythagoras said they should if they were to escape the Wheel of Birth.

Most disturbing of all, that boy Pythagoras had paid to do nothing kept bragging about his mysterious powers as if he were a wizard himself. What in the name of Zeus was a hypothenuse? And how could the square on the hypothenuse of a right-angled triangle be equal to the sum of the squares on the other two sides, if nobody knew what a hypothenuse was? The conceited young upstart told them it didn't matter whether anyone knew; he could "prove" all that rigmarole about the squares. When he proceeded to do so his unreceptive elders gathered that "proof" meant scratching a tangle of lines in the dust with a pointed stick. It looked like some new kind of magic. Probably it was. They decided that it must be magic of a very potent kind when the boy told them that Pythagoras had paid Apollo a hecatomb for this magnificent "theorem."

According to the enthusiastic pupil, Pythagoras had actually sacrificed a round hundred of prime cattle to his heavenly father when the latter divulged all the truth about the hypothenuse of any right-angled triangle. The Egyptians and Babylonians had urged Pythagoras to ask Apollo what the truth might be about any hypothenuse. They already knew what it was for a right-angled triangle with two equal sides; and some of them had even suspected the tremendous general truth when Apollo revealed it to his son. What was more, if the boy could be believed, it was Apollo himself who showed Pythagoras how to "prove" this grandest theorem in all "geometry." Now anybody with just a little sense could do the same. It was easy when you knew how. Possibly; but the elders doubted. In any event they were not going to have their bright boys turned into haughty young wizards under their very noses. This sort of thing must be stopped.

It was. The conservative element laid the facts in the case before the tyrant of Samos, their good friend and dictator of their minds. This able despot was shrewd enough to know that the only competitor he need fear was brains. Obviously this man Pythagoras was intelligent to a fault. The tyrant invited him to leave Samos.

At this critical juncture in his career Pythagoras hesitated. Should he submit to the tyrant and desert his own people? Or should he stay and share with them as much of his hard-gained enlightenment as they were capable of receiving? He knew that the tyrant, a petty opportunist with the mediocre mind of a practical politician, would be no match for him in a contest for the people's loyalty. He could win them to his side in a week, if he so willed; and they would hound their tyrant over the cliffs into the sea. In a pinch he could cow them with a trick or two of the childish magic he had learned from the Egyptian priests. That old one of changing a serpent into a rod, and back again into a serpent, alone would be enough to make the people his slaves. Should he go, or should he stay? Plainly his earthly father had not much longer to live. Soon his ageing mother would be the only personal tie binding him to Samos. Not quite the only one: there was this brilliant boy, his first disciple, who must be given his chance to become a geometer. Whatever the decision, his one convert to reason must not be abandoned.

To appreciate the choice Pythagoras made, we may contrast

his background in the sixth century B.C. against our own. It has been said that the white race is divided into two irreconcilable factions: those who regret that the French Revolution of 1789 ever started; those who regret that this democratic upheaval stopped before it had well started.

In the sixth century B.C. there were no machines to lighten the heavy work. Consequently slavery for the majority was a necessity if a minority was to live comfortably and have leisure, among other things, to think. Democracy as the French revolutionists imagined it might become, and as many since 1776 have believed it may become, was not even a philosopher's dream in the time of Pythagoras. There was demos, the mob, from which our word democracy is derived, and there was aristos, the best, from which we get our word aristocracy. Democracy means, literally, rule by the mob; aristocracy, rule by the best.

The slaves were not counted among the best in the sixth century B.C., any more than are all our brainless machines included in our aristocracy, though it may be true that they rule us. It was as natural for Pythagoras to ignore ninety-five —or more—out of every hundred of his fellow Samians in reaching his decision as it would be for us, in a similar crisis, to make up our minds without consulting our machines. Slaves and machines, the one at highest subhuman to the philosophic mind, the other without souls of any kind to the scientific eye, would be taken for granted as commonplace necessities for the good life.

The best of his own people had rejected Pythagoras. Though their fear was an unwitting tribute of respect, they probably were beyond lasting enlightenment. But he had recently heard of a Doric colony at Croton in Southern Italy, governed by true Greek aristocrats. They would welcome the new wisdom. The light Pythagoras hoped to kindle in Croton would illuminate the whole world.

In making his decision Pythagoras had in mind both science, including mathematics, and his secret number magic. These were to be the new light of the world. The farthest-reaching consequence of his choice he could not possibly have foreseen. Nor would he have understood if it had been shown to him. The slavery which made it possible for him to live and think, and which he barely noticed, was to be ameliorated, if not abolished, by the natural growth of that part of all his knowledge and wisdom which he deemed the lesser. What he considered the greater part was to obstruct enlightenment and foster superstition.

Accompanied only by his mother and his disciple, Pythagoras sailed away from Samos. He had made the third and last slight turn in his course. His path now lay straight before him to his glory and his doom.

CHAPTER 11

Discord and Harmony

CROTON was ripe for Pythagoras. The colony had just suffered a humiliating defeat at the hands of Locris. But the dissolute, luxury-loving Sybaris offered the Crotonites a tempting prospect for easy prosperity. Because the Sybarites—whose name survives as a synonym for high living and low thinking—went in for the more esoteric forms of dissipation with discriminating gusto, therefore the Crotonites would cultivate the simple life. Indeed the voracious Locrians had left them little else to cultivate.

By rigid self-discipline and devotion to things of the mind, with due attention to athletics, the Crotonites hoped to recoup their losses at some not too distant date. While waiting for the male toddlers to mature into tough hoplites, the governing class would encourage the manlier virtues and keep a vigilant eye on the inevitable and progressive degeneration of the Sybarites. When their flabby friends were no longer capable of putting up a stiff fight, the hardy young warriors of a disciplined Croton would fall on them and erase them from the surface of the earth. Such was the smoldering volcano into which the unworldly Pythagoras ventured, to propagate his gospel of enlightenment for all mankind.

The sage of Samos found an eager welcome waiting him. Croton had heard more than rumors of his austere way of life and of his powers as a magician. Here was just the man to unite the squabbling factions of the ruling aristocrats into a purposeful body with but a single thought, the annihilation of Sybaris. As events were shortly to prove, the aristocrats had reckoned without their general. Pythagoras was no drill sergeant to take orders from his superiors, for the sufficient reason that he had no superior. He would lead. They might follow, provided they had the necessary spark of intelligence.

The leader of the aristocrats was the world-renowned athlete Milo. This most muscular man in history, with the possible exception of Samson, was also the richest man in Croton and owner of the colony's most pretentious house. Not that anyone in Croton could be styled wealthy at the moment; still, Milo and his family had enough to eat, with some to spare for an abstemious guest. At the final meeting of the committee on reception and welcome, it was resolved that Milo should furnish Pythagoras with room and board gratis for as long as he might deign to honor Croton with his distinguished—and, to be hoped, profitable—presence.

Pythagoras readily accepted. In fact he felt highly complimented, for Milo was more famous than Pythagoras. Twelve times victor at the Olympic and Pythian games, Milo held the all-time record for these heroic contests. On one unforgettable occasion at Olympia this champion strong man raced into the stadium with a live bull draped about his shoulders. Thus adorned, he paraded before the wildly applauding spectators for one hour and a half. Afterwards, it is said, he killed the bull with a single smack of his open hand and ate the whole carcass in one day. But this last sounds more like an athletic exaggeration than a historic truth. It was indeed fortunate for Pythagoras that his appetite was only moderate. The Milo household consisted of the robust athlete, his dutiful wife, and his comely, not too submissive, young daughter Theano.

Thus it came about that the strongest body and the strongest mind in all Greece were allied in the stupendous task of salvaging Croton from the bog of depression into which it had been tossed by the joyous Locrians. If the besotted Sybarites had had any foresight or intelligence at all, they would have drafted the entire male population and begun drilling immediately. And if Milo and Pythagoras had been about thirty centuries ahead of the social conscience of their age, they would have taken the slaves and the common people into their full confidence. Instead, like the best men of their time which they were, the two grand strategists of victory regarded all of the population except the governing class as a commodity to be shoved about and disposed of as they should see fit. If there is such a thing as fate, it must have permitted itself a sardonic smile over what it had prepared for Milo and Pythagoras no less than for Sybaris.

In the leisurely nineteenth century of Queen Victoria, a writer would sometimes step from behind his ambush of the impersonal "we" and square up to his quarry, the "gentle reader," face to face. This enabled the writer, not too immodestly, to point out some exceptional beauty of the story as far as it had progressed, and to hang onto the gentle reader for at least half a page, beguiling her—it was usually a lady—

meanwhile with seductive promises of greater beauties to come. For the first and only time in this book I now revert to this admirable practice. I shall maintain it throughout this section. I do this because I am about to perpetrate a deliberate swindle on you, man or woman; and I want you to be forewarned that you are about to be taken in exactly as were thousands of readers before you for many, many centuries.

It is a rigid rule of the writing game that a writer shall not let his reader down. If you are told that a swindle is coming, it is fair enough, and you may be amused to detect exactly where the swindle is. To play the game a little better than fairly, I may say it is in the next legend I transcribe about Pythagoras. This legend concerns the most important thing Pythagoras did for science, and one of the two most important things he—or any man that ever lived—did for our scientific civilization. The other was his development of mathematics as a deductive science.

On its face this legend is as good history as any undocumented piece of evidence that has come down from ancient times. It was accepted by many historians of science and philosophy for over two thousand years as a not too improbable account of what might actually have happened. Please note that I did not say this legend was accepted by scientists. To give you another hint, no scientist would "accept" this legend. A scientist would do something about it.

Perhaps you have read a book on "how to read a book," or "how to read two books," or even, if you have been morbidly sensitive about your mind's salvation, "how to read a page." In addition to the excellent instructions in all such practical manuals, it is well, when reading a book with anything in it about science, to use your own head. I see that I have almost given you the game; so without further preliminaries I pass to the swindle itself. What follows now is it.

Probably it was in Croton that Pythagoras discovered the physical facts on which acoustics and the arithmetic of musical harmony are based. Passing a blacksmith's shop one day Pythagoras was arrested by the clang of the hammers swung by four slaves pounding a piece of red-hot iron in succession. All but one of the hammers clanged in harmony. Investigating, he found that the differences in pitch of the four sounds were due to corresponding differences in the weights of the hammers. Without much difficulty he persuaded the blacksmith to lend him the hammers for two hours. In that brief time he was to deflect the course of western civilization toward a new and unimagined goal. With the hammers over his shoulder, he hurried back to Milo's house. There, to the fearful astonishment of the bewildered athlete and his wife, he immediately prepared the first recorded deliberately planned scientific experiment in history.

To each of four strings, all of the same length and of the same thickness, he attached one of the hammers. He next weighed each of the hammers as accurately as he could. How he did this does not matter; he did it. He then hung up the hammers so that the four strings under tension were all of the same length. On plucking the strings, he observed that the sounds emitted corresponded to those made by the hammers striking the anvil. By sticking a small lump of clay on the hammer responsible for the dissonance, he brought the note emitted by its string into harmony with the other three. The four notes, now perfectly harmonized, trembled forth on the air in a melodious chord.

Pythagoras was even more deeply affected than his awestruck audience of two. For in that mysterious chord he recognized the first celestial notes of the elusive music of the spheres which had haunted his dreams since he was a boy. As he knew the weights of the hammers—they should have been perfect globes of pure gold—he quickly inferred the law of musical intervals. To his astonishment he discovered that musical sounds and whole numbers are simply related—how, is immaterial for the moment. It was a great and unprecedented discovery, the first hint that the laws of nature may be written in numbers.

There it is. What is wrong with it? If you studied physics in school, you are disqualified from answering, because you knew before you read the story. But if you have never thought about sound and the physics of music, the test is severe. Did you imagine yourself duplicating the experiment Pythagoras is said to have performed? If you did, you get half credit. If you rigged up an experiment of your own, or if you tried banging on a bucket or a dishpan with metal implements of different weights, you get full credit. If you did nothing of the kind, try it the next chance you have. A good ringing wine glass with spoons, knives, and forks of assorted sizes offer an excellent opportunity for rousing your dinner partner to the wonders of science when she is about to expire of boredom. (This suggestion is purely philosophic; I have never tested it by experiment.) You will find that the weights of your "hammers" have nothing to do with the tone. And so at any time in all the centuries from Pythagoras to Galileo, two minutes in a blacksmith's shop would have convinced the erudite scholars, who kept passing this absurd legend on to their successors, that the whole story is physically preposterous.

To conclude this Victorian digression in the classical manner, I append its moral. This is extremely important for what was to be the future of science when Pythagoras performed the first recorded physical experiment in history—not the farcical experiment of the legend, but a real and much simpler one. The moral is this: instead of emulating Pythagoras by resorting to experiment to find out what the facts might be, all but a very few of his successors were content to read about what he was said to have done. They did not appeal directly to nature to ascertain the facts of nature. They either cited authority or trusted their very fallible imaginations. The modern scientific age, that might have begun with Pythagoras in the sixth century B.C., was postponed by this physical and mental inertia till the late sixteenth century.

Modern science began when Galileo experimented with falling bodies instead of taking Aristotle's word for what "should" happen, but which does not happen. To a sedentary philosopher it is the most natural thing in the universe that an iron ball should strike the ground before a wooden ball of the same size if both are dropped from the same height at the same time. Try it, as Galileo did, unless you already know by actual experience that they strike the ground simultaneously.

There is another historical detail of capital importance. One of the earlier steps of the life sciences is classification. It is convenient, for instance, to have the plants and the animals all neatly parceled into families according to certain features common to all the members of a particular family.

But this step does not take us even over the threshold of the physical sciences. There more than passive observation is necessary. To discover anything useful about optics, say, light must be observed under artificial, man-made conditions that never occur in free nature. For example, it is observed before and after passing through prisms and trains of lenses of various curvatures, and even through the air between the poles of a strong magnet. In all of these humanly contrived situations anything that is measurable is measured as accurately as possible. Thus numbers enter the description of physical phenomena, and the "laws" of physics are expressed, as far as feasible, numerically.

This purposeful, deliberately planned interference with nature in the raw is what distinguishes the experimental-mathematical method of modern science, initiated by Galileo, from the earlier method of passive observation and classification. The earliest recorded practitioner of active interference with natural phenomena was Pythagoras. It was partly his own fault that worthy successors were so long in coming. We shall see later how he himself retarded his full greatness.

I now relinquish the gentle reader's hand, but not, I trust, the gentle reader's attention. What is to follow immediately marks the point in time where scientific man parted company forever with primitive man, and a new dimension was added to human thought. Without that addition our own civilization might be no more advanced materially and technically than the dead civilizations of Egypt and Babylon.

If one landmark overtops all others in the evolution of science, it is the discovery by Pythagoras of the connection between musical harmonies and numbers. He found that the notes emitted by vibrating strings depend in a very simple way only on the lengths of the strings, provided the strings be all of the same kind and all under the same tension. In particular he observed that the lengths which give a note, its fifth, and its octave are in the ratios of 6 to 4 to 3 or, what is the same, in the ratios of 1 to ⅔ to ½. An alternative statement of this is that the fifth and the octave of a note can be produced on one string under tension by "stopping" the string at ⅔ of its length, for the fifth, and at ½ of its length for the octave. From this epochal discovery Pythagoras proceeded to the construction of a diatonic scale. This implied much of orthodox music for many centuries. It also precipitated a great deal more, including the golden age of number mysticism almost immediately, and a delayed faith in deliberately planned experiment as the most profitable approach to nature.

A believable account of how Pythagoras made this decisive discovery credits him with the invention of the monochord. It was by experiments with this simple scientific apparatus that he found the wholly unexpected correlation between certain musical intervals and whole numbers. The apparatus consisted of a single string stretched on a board, with a movable wedge or "bridge" (like the bridge on a violin, but not fixed) between the string and the board. By moving the bridge the tensed string could be readily separated into two segments either of which might be made to vibrate independently of the other. The tensions of the segments remained (very approximately) constant as the bridge was moved to positions ½, ⅔, ¾, etc., of the string's entire length, and the lengths of the vibrating segments could be accurately measured.

An even simpler apparatus, which any savage might have made as far back as the Stone Age, would have sufficed. A heavy stone suspended from a branch of a tree by a thong of reindeer hide would have done. Even without a consciously planned experiment such as this, the Pythagorean law of musical intervals might have been discovered.

Primitive hunters and warriors without number must have noticed the twang of a bowstring. How many of them did more than merely notice? If any did, they left no mark on civilization. Going infinitely beyond static observation, Pythagoras interfered with nature, and by that dynamic act brought something new into the world, the art of scientific experiment. So far as is known, he was the first to think of making any apparatus at all with the deliberate intention of using it to force nature to answer a definite question: is harmony connected with number, and if it is, what is the precise connection?

It is small wonder that legend makes Pythagoras the son of Apollo, god of music and song. Even a modern scientist must marvel at the sheer luck which prompted Pythagoras to choose a promising subject for experimental investigation. With innumerable phenomena all around him to stimulate his curiosity and provoke his restless imagination, the master selected the one scientific problem which of all was ideal for a speculative mathematician. The electric sparks from rubbed amber must have mystified him as they had his teacher Thales; but Apollo or his own scientific instinct deftly steered him away from that intricate mystery. Had Pythagoras sought number in electricity he might be seeking yet. And so might we; for many simpler facts of nature had to be understood before electricity was approachable, and the necessary understanding came only by patiently following the way of experiment first pointed out by Pythagoras. Not until the twentieth century were the units of electricity isolated, when it was shown experimentally that electricity conforms to the Pythagorean dream of whole numbers.

But in acoustics the search was short. The relation between numbers and musical intervals is almost on the surface of physics and requires only the simplest apparatus to uncover it completely. And what was equally fortunate for the first experimenter, the relationship involves in its most immediate aspects only the simplest kind of numbers, the positive whole numbers 1, 2, 3, 4, . . . and their most obvious ratios ½, ⅔, ¾, . . . So if by being first in the art of scientific experiment Pythagoras was one of the greatest scientists in history, he also was one of the luckiest. Something impelled him to choose the one physical problem of all those crowding on his attention which he might have the faintest hope of solving. His happy choice may have been only blind luck. Yet anyone with normal senses can find innumerable problems worth attacking, and luck favors only those who not only are prepared to recognize it but who also understand themselves—"Know yourself" was the counsel of Thales. It takes genius of a high and rare order to recognize which problems worth solving are within one's powers and which are not.

It is sometimes said that Pythagoras did nothing fundamentally new, because exact observation in astronomy was already an old story when he was born. This misses the crux of the matter completely. In astronomy we observe, record our observations, reduce them whenever possible to numerical statements, and frame hypotheses to correlate what we observe. If a hypothesis fails to accord with further observations, we cannot find out how to modify it by performing a terrestrial experiment. We may refine or change our methods of observation and calculation; but that is a radically different thing from controlling to some degree the phenomena we wish to observe. There is no way of getting at the heavenly bodies to shift them about and, at will, vary the conditions under which they are observed. We may only look on; we cannot interfere. But in science of the kind initiated by Pythagoras the observer can control the conditions under which things are observed. If variations of temperature, for instance, are disturbing our precise measurement of a metal rod, we can easily keep our apparatus at a constant temperature. But nobody yet has succeeded in abolishing all but two of the heavenly bodies in order to simplify the problem of accurately describing the motions of the planets. In science as first practiced by Pythagoras the new and decisive element of purposive interference with nature in the raw enters. He might have listened to the harmonies of nature till he was old and deaf and have been no wiser than his incurious ancestors in the Stone Age. But when he began stretching strings, plucking them, and measuring their lengths, he endowed science with a new sense.

It will have been noticed that in all this the observer is indissolubly bound up with what he observes. Experiment and experimenter cannot be separated. How much of what the experimenter observes and measures is in nature, and how much in himself or in his methods of observation and measurement? The question is ultimately of the same kind as the one about the invention or the discovery of numbers. Pythagoras does not seem to have been troubled by it; Plato apparently was. But it was not until the twentieth century that the metaphysics of the physical sciences raised this question in an acute form.

At one extreme are those who believe that the only meaning of a particular thing in physics is the sequence of operations which are performed in observing that thing. For these "operationists" it is meaningless to ask what the "Reality" behind the observations and operations may be. At the opposite extreme some modern numerologists appear to believe that nature is forever humanly unknowable. All that we have imagined we knew about nature is merely what we ourselves have put into nature.

The extremists meet in a common ignoramibus: science as now practiced is incapable of telling us what life is. The simplest kind of experimenting with living tissue, for example, mere examination of it under a microscope, alters the tissue. What we hoped to examine, life as it is without mechanical, optical, or other interference, is no longer a datum of our experiment. So here is one region of human inquiry where experiment does not answer all. Like the capital question about numbers, "What is life?" may seem meaningless or improperly posed to our successors. But no such doubt chilled the enthusiasm of Pythagoras in the first hot flush of discovery. His law of musical intervals gave him the meaning of life itself. If not actually a number, life for Pythagoras was a shadowy manifestation of number. In some sense everything was number. That was his simple, all-inclusive solution of the universe.

Who can blame the enthusiast for overstepping the line between verifiable fact and unverifiable speculation? Such a discovery as that of the law of musical intervals might well have astounded and elated any man. Its total unexpectedness can be felt even today. Who would suspect that space, number, and sound are combined in one harmony? Space entered the close combination with the length of the plucked string, number with the ratios corresponding to the musical intervals. Sounds are distinguished by the ear; what has hearing got to do with numbers? And more unexpectedly yet, why should certain simple ratios of whole numbers have any connection with harmony, which is a province of aesthetics? All of these superficially unrelated things were not separate and distinct as they appeared. They were manifestations of one deep underlying reality. What could this ultimate reality be? Pythagoras disposed of all doubts by declaring that "Everything is number."

In the blinding light of this sudden revelation Pythagoras and his dazzled disciples lost sight of the experimental method responsible for the revelation. Turning away from the method that would have brought a scientific civilization within their reach, they followed the pure abstraction of number to its last fantastically barren refinement in an impossible numerology. Experimental physical science in the consciously questioning spirit of Pythagoras was not significantly resumed till the late sixteenth century, when Galileo went on where Pythagoras had left off. Roger Bacon, in the thirteenth century, and a few others before him, had made sporadic attempts to reintroduce the experimental method into a scandalously verbalized science; but Galileo was the first to succeed consistently and the leader whom a great and constantly growing host followed. He and the scientific fraction of Pythagoras were two of a kind, and to these men, more than to any other pair in history, western civilization is indebted for what it is today.

No sooner does a great philosopher solve the universe than a greater philosopher unsolves the solution. Occasionally the solver himself detects the fatal flaw in his solution. He then may do one of three things: admit that he was mistaken;

modify his solution so as to avoid the flaw; try to suppress the destructive discovery.

Though it is hard to believe, Pythagoras is said by some to have chosen the only dishonorable possibility of the three, when he learned that not everything is number in the sense he intended. Fortunately for the master's admirers, the legendary histories are so confused and mutually contradictory on this critical point that they effectively cancel one another. Perhaps at this distance in time it does not matter what Pythagoras did when he chanced upon the irresolvable discord that destroyed his numerical harmony of the universe. The important thing for science, mathematics, and philosophy is that his grand generalization was destroyed. The destruction of the Pythagorean "Everything is number," in the sense in which it was meant, was a major revolution in the development of all three disciplines.

By "numbers" Pythagoras meant the common whole numbers and the fractions or "ratios" obtained by dividing one whole number by another, such as $3/4$, $11/9$, $6/25$, etc. All of these, whole numbers and fractions, are called the rational numbers. These numbers were the only ones that had been invented or discovered when Pythagoras asserted that everything is number. It followed from the grand generalization that both a side and a diagonal of any square are measurable by (rational) numbers. But it was soon proved that if a side of a square is measured by a (rational) number, a diagonal of the same square is not measurable by any (rational) number. This destroyed the infinitely too simple generalization that everything is number.

The fact about the diagonal is phrased today as "the square root of two is an 'irrational number.' " Where was the square root of two before it was recognized by the Pythagoreans? Did that "number" exist in the nature of things, only to be observed by Pythagoras or his disciples? Or was it invented by the great mathematicians who followed Pythagoras? These men, especially Eudoxus (who flourished about 370 B.C.), developed a mathematical theory of "magnitudes"—such as lengths, areas, volumes—that was capable of strict reasoning about the "magnitudes" required to measure any finite length. The irrationality of the square root of two was phrased as "a diagonal and a side of a square have no common measure." But to perfect their logic, they and their successors were forced out of the mathematical finite into the mathematical infinite, and from the countable to the uncountable. Was that logic discovered or was it invented? And was the infinite a human invention, or was it a discovery by human beings of something which had existed before our planet cooled sufficiently to support animal life, and which will continue to exist when the earth is a dead world?

Whatever may be the answers to these questions—if they are indeed answerable and not pseudo-questions devoid of meaning—one thing is indisputable: the discovery about a diagonal of a square with rational sides was fatal for the simple generalization which had reduced the universe to rational numbers. In the numerical sense, the universe was seen to be irrational. (It is unfortunate that "irrational" has two common meanings, both relevant in discussions of the Pythagorean philosophy. When "irrational" is used in the sense of "contrary to reason," it will be unqualified; when "irrational" refers to numbers, it will be explicated as "numerically irrational.") It was not till our own time that a reputable scientist was unorthodox enough to doubt the rationality of the "laws" of nature. This will be noted when we recount that last great dream of Pythagoras, in which he passed through a hell of his own imagining. Here it is sufficient to observe that such a doubt is implicit in the questions about the rationality of the logic devised to rationalize the numerical irrationality of certain numbers.

It was proved in the late nineteenth century that if irrational numbers exist or can be created, they are infinitely more numerous than rational numbers. But this catastrophic dethronement of the numerically rational had no greater effect on modern numerology than had the comparatively mild revolution of the sixth century B.C. on the numerology of the Pythagoreans. They and their successors continued to philosophize as if the universe were a numerically rational creation of the common whole numbers. Experiment was impotent against anything the numerologists might claim. Number mysticism began and ended in the intangibles of the mind. It was beyond any objective scientific test, and it still is. That may be the secret of its apparent indestructibility.

It remains to indicate the part Pythagoras may have had in the subversive discovery. Some recognized authorities on the history of Greek mathematics believe there is no reason to doubt that Pythagoras himself made the fatal discovery, and this is backed by ancient traditions. Of the legends that may be accepted or rejected as we please, some state that when Pythagoras made the discovery he swore the members of his Brotherhood to secrecy. One account embroiders this by adding that the unruly brother who divulged the awful secret to the unhallowed mob was drowned. This sounds like pure fable; for what would be the point of drowning the man after he had published the truth? Moreover, the Pythagoreans were averse to the taking of life, human or other.

On the whole it seems not unreasonable to believe that Pythagoras took the unwelcome discovery in his stride and proceeded majestically on his way through space, number, and time as if nothing disagreeable had happened. In any event, he and his brothers and sisters in the zealous pursuit of knowledge and wisdom through numbers continued to live in peace and harmony in Croton, while the Sybarites reveled themselves into almost total military incompetence. Understanding nothing about numbers or metaphysics, Milo nevertheless was content to let Pythagoras teach these dark mysteries to his fellow aristocrats. He may even have discovered that there was nothing like a good stiff dose of arithmetic for keeping political meddlers so worried that they had no foolishness left to annoy the army.

CHAPTER - - - - - - - - - - - - - - 12

Harmony and Discord

TWENTY-THREE years slipped by so quietly in peaceful Croton that Pythagoras and his disciples scarcely noticed their passing. While Milo and his captains drilled the youth in the rigors of military discipline, Pythagoras marched and counter-marched his devoted followers through all the empires of the mind. They too were a well-disciplined company.

Only those who by severe deprivations had proved themselves capable of self-restraint and sustained thought were accepted as full members of the Pythagorean Brotherhood. Neither high birth nor influential office in the community sufficed of itself to admit an applicant to the master's lectures. Aspirants lacking the minimum requirements of rugged intelligence and ascetic disposition were impartially blackballed and rigidly excluded. Women were admitted under the same conditions as men, probably an unprecedented liberality in the sixth century B.C. There were two grades of membership, listener and mathematician. A sufficiently intelligent listener might graduate into the select circle of the mathematicians and become a full member of the Brotherhood, with a voice in the determination of policies.

First and last the tone of the organization was aristocratic.

The Brotherhood's exclusiveness undoubtedly assured a high standard of intellectual attainment among its members. But it also earned the devoted seekers after "truth in numbers" the cordial dislike of the common people of Croton and of those aristocrats who had been blackballed. One of the latter in particular, an aggressive malcontent by the name of Cylon, accepted his rejection by Pythagoras with marked ill-grace. In fact he dedicated himself to revenge.

Cylon had been a listener, but had lacked the necessary nous to become a mathematician. It is interesting while comparing the histories of the Pythagorean Brotherhood, compiled at different epochs, to note the fluctuating estimates of Cylon's character. When the historian is a Tory writing for Tories, Cylon is an unscrupulous demagogue.

A democratic historian addressing himself to his fellow citizens presents Cylon as a champion of the people and an advocate of equal opportunities for all, in short, a democrat. We shall simply tell in the proper place what happened. Cylon will not reappear until almost the end of the story. But the fact that he did nothing spectacular in the interval between his blackballing and his erasure of that humiliation does not imply that he was wholly idle. The poet who averred that "Hell holds no fury like a woman scorned" evidently had never met a prominent citizen of a small town who had been snubbed by the community's most exclusive club.

Estimates of the Brotherhood founded by Pythagoras are as various as the estimators. All agree, however, on the historical fact that the influence of the Pythagorean Brotherhood on mathematics, science, numerology, and philosophy was profound and enduring. It has lasted to the present day. On the social side, the degrees, the rites accompanying initiation and elevation from the grade of listener to that of mathematician, the rigid secrecy sworn to in an age when oaths were observed with superstitious reverence, the jealously guarded mysteries —all this and more of the same general character motivated by exclusiveness—fixed the pattern of secret societies for hundreds of years. The irrepressible tendency to oriental mysticism in the master's maturer teachings attracted the weary and the disillusioned, who longed to escape from the brutalities of a competitive world into a monastic peace where their wills were no longer at their disposal and where every decision was made for them. So rich indeed was the Brotherhood in regulated escapism that it served for centuries as a mine from which innumerable cults drew all they desired of ritual and creed.

A few details will suffice to indicate the kind of life the Pythagoreans lived and the rigors of the discipline to which they submitted. The harshness of a listener's probationary period was extreme. For three inhospitable years the would-be mathematician was hazed unmercifully. Should he venture an opinion or offer a harmless remark, his seniors first rudely contradicted him, then smothered him in ridicule and contempt. If the candidate was worthy, a year of such browbeating was usually enough to inculcate the virtues of silence and forbearance.

A meager diet, with no animal food except a scrap now and then left over from sacrifices to the insatiable gods, enforced the lesson of moderation. The generous wine which cheered the common man was prohibited, except for a sip or two before going to bed purely as health insurance. Any tendency to gourmandizing was checked by seating the patient comfortably at a loaded banquet table, letting him savor the appetizing aromas in anticipatory ecstasy till he reached for his favorite dish, when it was snatched away. His garments were scanty and coarse, but sufficient to keep him rugged. Even the solace of oblivion was denied him until he learned to get along on three or four hours' sleep and like it. Any little comforts he might have brought with him to soften his purgatory followed all his more substantial possessions into the common stock, and he enjoyed them no more. But if the discipline proved too severe and he resigned, everything was restored to him, and he was discharged with no further obligation than a promise to keep what he had learned to himself. Cylon succeeded in his revenge partly because he broke his promised silence. When a candidate finally got used to the life he found it not much harsher than the basic training in a barbarous military camp.

While the body was being toughened the mind was by no means neglected. Long before sunrise the day began with semi-religious exercises. Lofty metaphysical poetry and elevating mathematical music hardened the auditors for a solitary walk of meditation before their cheerless breakfast. During this walk each planned his day. Good intentions were balanced against performance at sunset. Should some unhappy wretch do some things which he ought not to have done, or leave undone some things which he ought to have done, he penalized himself appropriately the next day.

The morning bread and water was followed by a short period of relaxation to prepare for the real rigors of the day. All gathered for a friendly chat. The few who had earned the privilege of expressing their minds spoke softly and sparingly while the others listened and said nothing at all. This unilateral style of conversation was designed to further the capital ideal of producing submissive minds in disciplined bodies.

The Pythagoreans were among the earlier discoverers of the physiological fact that hard physical work is a slow poison destructive of creative thinking. Being relieved by their slaves of the necessity of indulging in brutalizing labor, they kept themselves fit by judicious doses of cultural athletics. Wrestling bouts, running, javelin tossing, and similar sports sharpened their appetites for the tasteless evening meal of bread, honey, and water. As already noted, the neophytes were permitted a little wine. The mathematicians, supposed to be above such frailities of the flesh, got only pure cold water and not too much of that.

Any mathematicians still awake after their unexciting repast—consumed in silence—turned to the administration of the Brotherhood's domestic and foreign affairs. The survivors of this tedious ordeal refreshed themselves with protracted religious exercises of mystical solemnity, took a cold bath, and fell upon their stony beds. Up again some hours before daybreak, they plunged once more into the endless round of music, meditation, talking or listening, solitary promenades, introspection, unappetizing meals, numerology, science, mathematics, religion, athletics, metaphysics, bathing, and just enough sleep to prevent them from dozing off on their feet. It was no life for a sybarite.

At the peak of the Brotherhood's prosperity some two hundred families (other estimates give three times as many) lived together more or less harmoniously under the fatherly supervision of Pythagoras. As for the master himself, he enjoyed every moment of his undisputed authority. Numbers were not the only mysteries he understood better than any of his disciples. In cultist psychology he still is without an equal in all the long and varied history of cults. Always aloof, even when conferring with his brother mathematicians, he seldom spoke unless he had something mystifying to communicate. Taciturnity seems to have been a passion with him, if not for himself, then certainly for his followers. To ensure a properly respectful acceptance of his teachings, he imposed a silence of from three to five years on listeners newly promoted to the grade of mathematician. His disciples seldom saw him, but when they did they were overwhelmed by the majesty of his bearing. Like the master of showmanship he was, Pythagoras always chose the unexpected moment to exhibit himself. His rare appearances were rendered sufficiently godlike and remote by a voluminous white robe, a crown of golden leaves, and his full white beard. To heighten the mystery of his more recondite doctrines he intoned his most confidential utterances behind a curtain. The organ voice, accompanied by melodious chords struck out with bold abandon on his lyre, convinced the more credulous of his auditors that they were hearing Apollo. Pythagoras never made the mistake of stepping from behind the curtain when the last note of his musical discourse had perished in the quivering silence.

When the curtain began to wear threadbare the master retired with his lyre to the Grotto of Proserpine. Like others of the ancient oracles, Pythagoras knew by experience that the rumbling echoes of a human voice rolling from a gloomy and sulphurous cavern are irresistibly impressive to an uncritically receptive mind. For descending to such rather shoddy tricks of pedagogy Pythagoras has been called a charlatan.

He was not. So unquestioning was his belief in his message for his fellow men that he used any and all means within his power to get it accepted. He may have even convinced himself that the voice issuing from the cave was not his own but Apollo's. If so, it would not be the first time or the last that a great teacher elected himself the mouthpiece of divinity.

These details of the Brotherhood's way of life may be concluded with the story of a nameless brother which somehow rings true. The secret emblem of the Brotherhood was the mystic pentacle, the five-pointed star formed by extending the sides of a regular pentagon till they meet by pairs in the points of the star—like a star in the flag of the United States. One property of the pentacle which overawed the Pythagoreans is its unicursality: the star can be traced by the uninterrupted motion of a point without traversing any part of the star twice. A second property, profoundly numerological in character, mystified the mathematicians beyond all credence. The star has five points, and the Greek for health is a five-letter word. The five letters might therefore be affixed to the five points, one letter to each point. It followed numerologically that the unadorned, letterless pentacle must be health itself. The better mathematicians discovered numerous further properties of their pentacle and demonstrated all of them by the strictest deductive reasoning. Since the property just proved is the only one relevant to this story, I shall omit the rest.

A young and impecunious brother, so goes the legend, while traveling in foreign parts far from home fell dangerously ill. A charitable inn-keeper nursed him, although the young man had made it plain that he had neither money nor goods with which to settle his bill. When it became obvious that he was dying, the young man asked for a board on which to draw. He scrawled on it the mystic pentacle and told the inn-keeper to hang it up outside his door.

"Some day one who understands what I have drawn will pass by. He will stop and question you about the sign. Tell him everything, and say I ask him to pay you. You will be rewarded."

And thus it happened.

The prime concern of the brothers was to live what they considered the good life, and so to escape the more degrading turns of the Wheel of Birth. But being only human in spite of all their geometry and numerology, they could not refrain from seeking to extend their dominion from the immaterial to the material. On the earthier level the more exoteric practices of the brothers afforded a model for scientific academies and learned societies. Going far beyond the orbit of any scientific organization of our own day, the Brotherhood included statecraft and politics in its curriculum. Pythagoras taught that government should be by the best for the common good of the best—a sort of aristocratic communism. Many details of the Pythagorean theory of government passed almost unchanged into the ideal state advocated by Plato in his *Republic,* also into his *Laws.*

The only possible hitch in this eminently reasonable program was the selection of the best. Who was to do the choosing? The Pythagorean solution was as simple as it was final. At the apex of the government Pythagoras firmly placed himself. He then elevated the mathematicians almost to his own level. The "mathematicians" corresponded to the "guardians" in Plato's philosophically perfect society. Below the mathematicians the listeners supplied an appreciative but voiceless audience for their superiors. Below the listeners all but a small fraction of one per cent of the entire population constituted "the mob," that is, the major part of the body politic to which the Pythagorean theory of government was to be applied.

The principle of selection here seems to have been as sound—mathematically—as any human device possibly could be. For after a man had successfully passed the grueling discipline of becoming a Pythagorean mathematician he would certainly have mastered the rudiments of self-control, and ability to govern oneself was postulated to be a necessary and sufficient prerequisite for successfully governing others. From this it could be logically demonstrated that the theory must work in practice. Unfortunately for the Pythagoreans it did not.

It is of more than historical interest to note that one feature of the Pythagorean training in government passed unmodified into the educational program prescribed by Plato for his republican guardians. Not Pythagoras himself could have been more insistent than Plato on the value of mathematics in the training of future administrators.

But lest this dual endorsement of mathematics as a disciplinary preparation for statesmanship be urged too seriously today, we should remember what "mathematician" signified to the Pythagoreans. A mathematician was one who had survived several years of a merciless discipline and who, in addition, believed he understood what the master meant when he asserted that everything is number. In Plato's educational policy the last was replaced by an intuitive faith in the value of mathematical reasoning as a preliminary step, and a rather humble one at that, toward any consistent thinking, including dialectic and the transcendental logic of Plato's Eternal Ideas as sublimated in his Ideal Numbers. So whatever may be the merits of exercises in mathematics as a preparation for the less tractable problems of human relations, neither Pythagoras nor Plato can be honestly cited as authority that a few lessons in common arithmetic or elementary geometry will transform a mediocre politician into a brilliant prime minister or an astute president. Still less is it likely that an expert numerologist—a Pythagorean or Platonic mathematician—would be a good helmsman for the Ship of State today, however competent he may have been when Plato invented both helmsman and ship.

Though the deciding vote on all questions of policy was cast by Pythagoras, it would be unjust to stigmatize the Brotherhood as a dictatorship. Actually the organization was far more complex. The brothers (and sisters), it is true, acknowledged but one superior and one master, Pythagoras. Whatever they discovered was voluntarily attributed to him. Thus in the realm of the mind he was despot. Scientific or other impersonal disputes among the brothers were invariably settled with devastating finality by citing the master as authority: "Himself said it"—"Ipse dixit."

This intellectual absolutism did not carry over to the Pythagorean theory and practice of government. Convinced of their unapproachable superiority, the Pythagoreans resented any authority above their own. Both secretly and openly they opposed tyrants wherever they found them. The parent organization in Croton became a training school for political saboteurs, whose flaming zeal for the master's gospel drove them out to torment all absolute rulers within reach by land or sea. And nearly everywhere they settled the missionaries established secret societies modeled on their great headquarters in Croton. Politically these tight little islands of aristocracy in a rising tide of democracy were pernicious to the general welfare and, in the end, fatal to the Pythagorean Brotherhood itself.

The resultant political upheavals that followed these missionaries wherever they went, toppling one tyrant after another from his seat, might be interpreted as a long overdue and inevitable upsurge of democracy. But it would be stretching the sequence of cause and effect beyond the breaking point to say that the Pythagoreans responsible for these popular uprisings were democratic lovers of all mankind. The intransigent Pythagoreans neither loved nor hated the majority of their fellow men, for the adequate reason that they knew nothing whatever about them. All but their own rigidly exclusive few, living without recourse to trade or productive bodily labor of any kind, were as foreign as ostriches to the self-centered devotees of pure thought. This comprehensive ignorance of the society on which they battened and theorized was to prove the Pythagoreans' undoing. While the philosophical mathematicians were wrangling among themselves over the abstract problem of the One and the Many, Cylon and others like him were preparing a rudely practical solution of the human problem of the many against the few.

All this sums up to the verdict of history on the Pythagorean Brotherhood. At the highest estimate the organization was a disciplined aristocracy of intellect devoted to the pursuit of impartial science and the furtherance of just government. The science was mostly mathematics and astronomy, of which more than a little was mystical or allegorical. The government was based on slavery and the assumed congenital inferiority of the mass of mankind. Slavery was accepted as a natural necessity and a just dispensation of the gods, and mass inferiority as a fact of common observation. As it usually is in aristocracies and democracies, justice was an absolute. At its basest the Brotherhood was a self-elected, self-perpetuating band of bloodless aristocrats dedicated to the preservation of their own special privileges by the exploitation of the mass of their fellow men.

Neither extreme gives a fair estimate of the Brotherhood —if we may credit a host of contradictory witnesses. The Pythagoreans were neither a society of inhumanly wise altruists sworn to seek truth and uphold justice, nor yet a selfish clique of smug and callous snobs. Their successes and failures were characteristic of the age in which they lived, and it may reasonably be doubted whether they, as limited human beings, could have done any better than they did with the chaotic materials at their command. Some of their successes will occupy us later. To balance the account of the Pythagorean Brotherhood we must now record one of its major blunders—if it was such—which persists to this day.

The Pythagorean error was of a kind which almost any self-perpetuating society of the best people in almost any community might be expected to embrace. It is implicit in the simile by which Pythagoras himself illuminated his philosophy of life. His spiritual successors in science, mathematics, and philosophy consider this the finest thing their master ever said. Likening all humanity to a concourse at the Olympic Games, Pythagoras said, "Men are of three kinds: the lowest come to the Games to buy and sell; the next higher to compete; the highest come simply to look on. So it is," he said, "with life. And," he continued, "the most cleansing of all purifications from the taints of many lives is the pursuit of knowledge for its own sake. Only the disinterested philosopher, the man who loves wisdom for itself, is fully liberated from the ever-turning Wheel of Birth. The soul can be purged of evil only by the true knowledge, which is pure science, and only by unselfishly following this unprofitable knowledge can the soul escape the miseries of successive incarnations. Suicide is no way out, for it incurs the severest of all penalties. The pure theory of numbers offers the quickest escape from life. It is the least profitable of all forms of human knowledge."

The lofty and ennobling tone of this utterance has been echoed and re-echoed all down the twenty-five centuries since Pythagoras first preached the gospel of science for science's sake. It was taken as only moral that the slaves do what work was necessary to liberate the philosopher in order that he might liberate his soul. But let that pass. It is of but little consequence for the deeper issues: this life is an evil from which the good and the intelligent will withdraw; science is merely an anodyne to dull the pain of living, and is the more effective the less useful it is.

This oriental pessimism in the Pythagorean outlook on life may have been already ancient when the Brotherhood adopted it as their own. The one supportable life was the last turned up by the Wheel of Birth, and was no life at all, but total extinction and eternal oblivion. The most depressing feature of this philosophy, the condemnation to successive reincarnations in more or less degraded forms of life persisted, as we have seen, even in the highly idealized immortality of Plato. The hopeless pessimism, with its consequent withdrawal from life as it must be lived by all but a saintly or calloused few, lasted through the Middle Ages, filtering thence into numerous creeds and cults of our own time.

It seems rather remarkable that the Pythagoreans should have promulgated this particular form of inhumanity. They lived securely, and as well as they liked, and except toward the end of their exclusive little aristocracy had no intimate experience of physical suffering. Pythagoras himself might have witnessed Asiatic squalor and cruelty. If he had, he may well have concluded that life is a business to be expedited with as little living as possible. It is perhaps not so remarkable then that the Pythagoreans and their successors should have found the best of all possible lives in the pursuit of the purest of pure mathematics.

As for the devotion to science for the sake of science alone, opinion is sharply divided, particularly among mathematicians. The Pythagorean creed has often been challenged, especially in Russia following the First World War. It was imagined by the prophets of the new order that the ultimate justification of science may be the common good of the human race rather than the increase of knowledge for its own sake.

The practical desirability of cultivating so-called pure science is not questioned. From ancient Egypt and Babylon to the present day it has been demonstrated that applied science advances but slowly or not at all when pure science is neglected. It is the motivation which is in doubt. Were the aristocratic Pythagoreans right in aiming to make their mathematics, say, as beautiful and as impractical as they could, in order that mathematics might increase as rapidly as possible? Or are the more proletarian scientists right in striving toward a general betterment of the whole race, regardless of what uncouth shape mathematics may assume in the process? Without some generally accepted standard of values in such matters these questions are unanswerable. But the historical consequences of the Pythagorean "numbers for the sake of numbers" are as clear as they are important.

The kind of arithmetic that was useful in Greek daily life was called logistike, or logistic. From Pythagoras to Plato, and after Plato to the end of the great period of Greek mathematics, logistike, if noticed at all by competent mathematicians, was treated by them with contempt or scornful indifference. It was fit only to be learned by slaves who did what keeping of accounts might be necessary. Consequently, the Greek alphabetic system of writing numbers—so childish that to describe it here would be a waste of time—remained almost static. Such modifications as were introduced resulted in a style of numeration that one competent historian and sympathetic critic of Greek mathematics has characterized as vile.

The useless kind of arithmetic, that which deals with the properties of numbers as such with no thought of any application either to science or to daily life, was called arithmetike. This was cultivated with occasional brilliance by the Pythagoreans and their successors, almost to the end of Greek mathematics. Arithmetike invariably was regarded as a discipline worthy of study by all true men and by all governors of men. Number mysticism, frequently degenerating into utterly senseless travesties of reason, was equally respected by the Pythagoreans and their successors in philosophy. When the social status and other-worldly outlook of the Pythagorean Brotherhood is remembered, this peculiar separation of arithmetic into the reputable and the disreputable is what might have been anticipated.

Happily for the progress of civilization in the large it is no longer shameful, or even undignified, for a pure mathematician to concern himself with improvements in the lowly though useful techniques of calculation, for these, after all, are matters of pure mathematics. The Pythagorean mathematician today draws the line between reputable and disreputable mathematics somewhere above electrical engineering and below the theory of relativity. As in the days of Pythagoras, the farther removed from practical applications a mathematical discipline is, the more highly it is esteemed by the spiritual descendants of the master. Yet even a rudimentary knowledge of the history of mathematics suffices to teach anyone capable of learning anything that much of the most beautiful and least useful pure mathematics has developed directly from problems in applied mathematics.

Harmony and the mysteries of numbers did not absorb the entire attention of Pythagoras during his sojourn in Croton. Nor was he personally quite the complete ascetic he should have been had he fully believed in all of his own teachings. Living under the same roof as Theano, the master would have been superhuman indeed had he remained indifferent to her quite exceptional charms. This is merely the materialistic explanation of his romance. But true disciples of the master favor a prettier and more spiritual version, in which the marriage of Pythagoras is presented as an act of humane self-sacrifice on his part.

Milo's daughter was not merely beautiful, according to the legend; she was also unusually intelligent. There was no difficulty about admitting her at once to the upper circle of listeners, where she soon proved herself the most attentive of the master's auditors. To her, if tradition can be credited, is ascribed the only biography of Pythagoras by one who knew him in the flesh. Unfortunately this work was early lost, though statements based on its alleged authority have survived. Theano's account of the master included his teachings as well as intimate details of his life, and is said to have been the inspiration of his immediate followers.

For years before she finally broke down and confessed her hopeless infatuation for her teacher, Theano was his favorite pupil. Pythagoras is pictured as being so deeply submerged in his numbers and mysticisms that he was shocked, surprised, and finally delighted when Theano informed him that she could stand her torment no longer and was about to expire of an unreciprocated passion. On persistent questioning by the master, she gave in at last and disclosed the name of the man whose love, according to the legend, she craved but felt herself unworthy to receive. It was Pythagoras. To save her sanity, if not her life, Pythagoras sacrificed his asceticism and married her. Theano's proposal and marriage took place in the Grotto of Proserpine, a singularly inappropriate locality for a courtship with its pointed suggestion of the Greek hell. But it was there that Theano sought and found the master she adored, and it was there that Pythagoras suffered his last dream.

Despite the considerable disparity in their ages—some accounts say as much as forty years—Pythagoras and Theano were happy in their marriage. It is claimed, how reliably seems not to be known, that Theano bore Pythagoras two sons and a daughter. One of the sons is said to have taught Empedocles (flourished, 450 B.C.), to whom he transmitted all the secrets of the master preserved in Theano's life of her husband. There may be a shade of truth in this, for Empedocles acquired a legendary reputation as a worker of miracles second only to that of Pythagoras. Empedocles followed so shortly after Pythagoras that many of his teachings differ but slightly from those attributed to the master himself. Some of these must be noted later. Though Empedocles was no fanatic for numbers, his philosophy is in the direct line from the numerology of Pythagoras to its climactic refinement in Plato's Ideal Numbers. More doubtful accounts make Empedocles a personal pupil of Pythagoras. Though unlikely, this direct contact with the master would not be chronologically impossible. However he acquired it, Empedocles transmitted the Pythagorean philosophy to others who passed it on to Plato.

The years of peace drew rapidly to a close. The Brotherhood's missionaries had propagated their gospel more effectively than they knew: democracy was stirring everywhere in the Greek world. By a curious irony it was an act of humane generosity on the part of Pythagoras himself that precipitated the disaster to the Brotherhood and ultimately his own downfall.

Decadent Sybaris had its political upheaval shortly before austere Croton experienced hers. A considerable number of the upper class had rashly disagreed with their tyrant. He, being the stronger, won the argument. Five hundred panicky Sybarites, the elite of a degenerated aristocracy, implored the government of Croton to grant them sanctuary. Death was at their heels, they said, and indeed they were not far wrong. Pythagoras convoked the council and laid the urgent petition of the refugees before them. Fearful for the integrity of their own skins, the aristocrats of the council denied the prayer of their brothers and sisters across the border. To have given the outcasts shelter might have incurred the displeasure of the Sybaritic democrats. It was then that Pythagoras showed what was in him. Overruling the council, he invited the refugees to come on. They came, precipitately. The democratic party of Sybaris, now in full control of their government, demanded that Croton extradite the refugees. Croton—in effect Pythagoras—refused, and Sybaris immediately declared war on Croton. Promptly and with savage joy Croton took up the insolent challenge.

Milo was ready. Leading his perfectly disciplined troops in person, he marched on the enemy's capital. His supple young warriors leapt on the flabby Sybarites, butchered their inefficient soldiers, their old men, their children, and their women—with the exception of a few earmarked (literally) for slavery. Then they demolished every house and hovel in sight and, after months of hard labor, diverted the river Krathis to bury the wreckage. This devastating blitz in the classical manner was possible only because the obliterated Sybarites had taught the Crotonites the simple virtues of abstinence and obedience. Their defeat at the hands of their unwilling pupils showed how effectively they had taught and paid them in full for their labor.

After victory the headache. Croton's was brought on by one of the more frequent causes: an equitable division of the spoils. Milo had seized all the lands that had formerly supported the ruling class of Sybaris in luxury. Whose were they now, the aristocratic Pythagorean Brotherhood's or the democratic mob's? The veterans who had done the bleeding began agitating for a bonus. They found their mouthpiece in

Cylon, now leader of Croton's democratic party. And Cylon, for his part, recognized in the clamor of the mob for land the opportunity he had nursed for all of twenty bitter years—ever since he had been expelled by the Brotherhood—to get even with "that pompous old snob and holy fraud Pythagoras." Thus he characterized the master to his howling constituents.

Serene as usual, Pythagoras ignored the clamor. Let the rabble hoot and yell; everything was still number. He and his brother mathematicians continued to work calmly, perfecting their solution of the universe. Unaware that the peaceful world they had loved lay shattered and mute, the Brothers proceeded with their proof that the Earth is a pure and flawless note in the celestial harmony of the spheres.

CHAPTER - - - - - - - - - - - - - - 13

Mythology Transformed

IT WILL be well at this point to take a quick glance ahead at the net outcome of the Pythagorean Brotherhood's teachings before inspecting a few of the curious and more important details that influenced rational thought for many —perhaps too many—centuries.

Before the Pythagoreans imagined they had reduced everything to numbers, mythologies of the universe were largely anthropomorphic. To account for storms, spirits of the wind, the thunder, and the lightning were invented, and so on all up or down the scale of natural phenomena. The Pythagoreans swept all these crude personifications away with their universal arithmetic of nature, substituting for "the lengthened shadows of men" as rulers of the universe purely abstract mathematical fictions. The aim remained unchanged: to give a rational picture of the world as it appears to human beings. The gods served their purpose well enough until rationalists like Thales, Anaximander, and Pythagoras suspected that impersonal reason might be more effective than theology in the representation of nature.

It was not the purpose of the new interpretation to nullify the old. The immortal gods were left in full possession of all their rights and privileges, including that of worship by any human beings who might still believe in their existence. Though the rationalists were frequently persecuted and occasionally killed by too zealous believers, very few of them sincerely doubted the reality of the gods. Most seem to have agreed with certain modern scientists and mathematicians that the natural and the supernatural can be consistently accommodated in one mind.

The rationalists thus differed from their more orthodox fellows only in their mild heresy that nature may be understandable through symbols less primitive than anthropomorphic deities. They then proceeded to elaborate their own symbolic representations of nature. These too were mythologies. When Thales asserted that everything is water, he was as much of a myth-maker as the nameless Indian who preserved the earth by planting it on the back of a turtle while the gods churned the oceans. But with the Pythagoreans nature myths began to suffer a radical change. They became progressively dehumanized and increasingly abstract. About two and a half centuries of etherealization were to reach their climax in the elusive Ideal Numbers in which Plato attempted to embody his Eternal Ideas. These Numbers then became the ultimate reality and the essence of all Being. Abstraction could go no farther. And just as their ancestors had satisfied themselves that they had explained everything once for all when they invented the immortal gods, so Pythagoras and Plato believed they had attained finality in their own rarefied myths.

These two master mythologists between them determined the subsequent course of speculation regarding the nature of the physical universe. About twenty centuries were required for the more scientific Pythagoreans to evolve at long last into classical mathematical physicists in the hard-headed, commonsensical tradition of Galileo and Newton. Almost none of these busy artisans ever thought to enquire what their labors might signify in a credible theory of knowledge.

Meanwhile the strictly mathematical Pythagoreans succeeding Plato were developing into classical pure mathematicians, the majority of whom adhered to their fundamental creed that numbers are revealed rather than invented. They also believed the like about the laws of classical logic, the theorems of geometry, and the gods.

Human arithmetic and geometry for the Platonic mathematician today, as twenty-three centuries ago, are imperfect descriptions of an ideal Arithmetic and an ideal Geometry, both superhumanly perfect, existing timelessly in a realm of Eternal Ideas forever inaccessible to direct human knowledge. What terrestrial arithmetic and geometry may reflect of this celestial Arithmetic and Geometry is but the blurred image of a Truth no mathematician will ever behold. Yet the outlook is not wholly discouraging. By selfless devotion to the pursuit of pure knowledge the soul of the mathematician is itself purified till, in it, as in a mirror of burnished silver, arithmetic and geometry appear in fleeting glimpses as Arithmetic and Geometry. Only when the mathematician's soul is completely liberated from his body may it reflect Arithmetic and Geometry clearly and flawlessly.

About the year 1920 a few descendants of the more scientific Pythagoreans joined their Platonic brothers in the realm of disembodied ideas. Until general relativity (1915) inspired the followers of Galileo and Newton to ask precisely how much of their "laws of nature" had been put into science by their own techniques of observation and experiment, and how much was inherent in nature and independent of all observers and experimenters, a large majority had believed they were describing nature "as it is." The few then began to question. Were their "laws of nature" natural after all, or were they merely trivial consequences of the manner in which all sane human beings reason? Relativity had made singularly unexpected predictions, subsequently verified by observation, by purely mathematical reasoning applied to truisms that seemed —after they had been pointed out—to be necessary for any consistent thinking about the physical universe.

By 1940 everything began once more to evaporate in numbers, but less violently than in the Pythagorean mythology of science. Twenty-five hundred years ago, a quarter of a century sufficed to arithmetize the entire universe as then known. An equal span from 1915 to 1940 compassed only the beginnings of a numerology of one science, physics. Compared to what the ancient Pythagoreans accomplished, the achievements of their modern rivals are as yet somewhat meager, though doubtless pregnant with infinite possibilities. It may help us to value the new as it merits if we now sample the old quite liberally.

As Empedocles had an important part in the transmission of Pythagorean numerology to Plato, we may consider him and his contribution to the Greek attempt at physical science first, though this puts him slightly ahead of his chronological place in the record. We have mentioned that Empedocles (flourished, 450 B.C.) may have been a pupil of Pythagoras. Whether or not he actually knew Pythagoras, he was a confirmed Pythagorean in his thoughts if not always in his deeds.

There seems to be but little doubt that Empedocles was

slightly off balance. The son of a very rich father, he inherited more wealth than he could possibly squander on himself, even while gratifying his somewhat expensive tastes. But Empedocles had an original mind, and the problem of what to do with his superfluous wealth caused him no difficulty. To ease himself of his burden the future philosopher conceived what must be the most bizarre scheme in history for getting rid of unwanted riches. He hunted up all the poor but otherwise desirable girls he could find in his native Acragas (later Agrigentum), forced handsome dowries on them, and married them off to the needy sons of the best aristocratic families in town. As Empedocles never took a wife himself, he may have done it all as a sardonic joke. Or he may only have been the pioneer eugenist.

In showmanship and dignity, not to say pomposity, Empedocles surpassed even his master Pythagoras. Purple being the hue of tyranny, the philosopher showed his contempt for tyrants by arraying himself in a shrieking purple robe. To enhance the impertinence he adorned his middle with a chain of pure gold and his head with a chaplet of golden leaves. And to hint that he could easily settle any dispute that might arise, he supported a larger and better fed retinue than any tyrant could afford. All this, with his own wisdom and eloquence thrown in, he placed at the disposal of his oppressed fellow countrymen though he himself was no democrat. In spite of his high seriousness, Empedocles must have had a wry sense of humor. When his democratic partisans overthrew the oligarchy of Acragas and implored Empedocles to be their king, he laughed them out of their well-intentioned stupidity and remained his own intractable self.

In his propagation of Pythagoreanism, Empedocles emphasized the health-cult feature of the Brotherhood, and himself performed prodigies of healing which have yet to be duplicated in modern medicine. It is reported that, surpassing young David, he permanently cured madness with music. He also restored a woman who had been dead thirty days to life. There may be some exaggeration here—possibly the patient was only in a deep coma. But the next, even if it never happened, is remarkable merely as an imagined anticipation of modern sanitation. Empedocles is credited with having wiped out malaria in a certain town by draining the surrounding marshes. As an engineer he bettered the record of Thales, providing his hot and humid city with complete air conditioning by cutting a pass through the mountains to admit the cool north wind. Medicine and engineering were only the most popular of his numerous accomplishments. Though his fame rests chiefly on his philosophy, Empedocles was also a considerable poet. He has some claim (though a doubtful one) to have composed the famous *Golden Verses of Pythagoras*.

It was inevitable that such a man in the fifth century B.C. should be acclaimed as divine, and tradition asserts that Empedocles did not disdain the compliment. Enthusiastic crowds trailed him wherever he went, inventing miracles for their god when none were forthcoming. If the wind veered from north to northeast, Empedocles had commanded it to do so; if it stopped raining, Empedocles had ordered the sun to shine; if the day was sultry, he had summoned the refreshing shower. His was a hard reputation to maintain.

Legend has it that Empedocles came to believe wholeheartedly in his divinity. To prove that he was a god, he dived into the flaming crater of Mount Aetna. Matthew Arnold's

Victorian version of this classic legend pictures Empedocles committing suicide, while temporarily of sound mind, to escape the senseless adulation of a persistent rabble. Like children at a circus the crowd clamored for novel tricks, and wondered audibly at the performer's gorgeous raiment when there was a lull in the healing of the sick or the taming of the winds. Empedocles was more interested in trying to teach them to govern their appetites and to appreciate the "elemental four" revealed in his philosophy. Failing, he destroyed himself. One of his brass sandals was subsequently recovered after an eruption. To his fickle admirers this was conclusive proof that Empedocles, though godlike, was not a god. A less credible legend pictures Empedocles ascending to Olympus in a blaze of glory—which may have been the eruption. A more pathetic account has him dying in lonely exile after expulsion from Acragas by his political enemies. His crowds did not follow him into exile. As with Pythagoras we may take our choice.

Empedocles is honored in the history of scientific numerology for his exploitation of the number four. To him is ascribed the theory of the four "elements"—earth, air, fire, water—which survived in Aristotelian science for far too many centuries. The four elements still haunt literary allusions. Not so long ago it really meant something to praise a well-balanced man by saying "The elements [were] so mixed up in him that Nature might stand up and say to all the world, 'This was a man!' " Today the connotation of "elements" would be merely ludicrous, for a man would be mixed indeed if he were composed of everything from hydrogen to transuranium. But in Shakespeare's day it was almost scientific, or at least not ridiculous, to mix him up out of earth, air, fire, and water. It was a suggestive metaphor, rich in the scientific associations of two thousand years of ossified Greek chemistry and cosmogony, and redolent of Plato's celestial arithmetic.

So deeply did Empedocles plant "four" as the metaphysical number of matter in the human mind that the labors of more than twenty centuries, three of them dominated by modern experimental science, were required to root it out. No number in all numerology has had a longer or more baneful pseudo-scientific career than this chemical four of the Pythagoreans. Its specious simplicity commended it to philosophers from Empedocles to Plato, who passed it on to generation after generation of uncritical disciples.

Though traditionally attributed to Empedocles, the four elements may not have been wholly his own invention. For it is now impossible to separate any particular contribution for which he, a devout and erudite Pythagorean, may have been personally responsible from the collective teachings of the Brotherhood. Not being under the direct tutelage of Pythagoras, however, Empedocles was not obligated to stamp all of his philosophy with the master's trademark, "Himself said it." Several of the details reported next passed, in but slightly altered form, into the science of Plato, who likewise deemed it unnecessary to attribute them explicitly to the Brotherhood. But he did put some of his most remarkable scientific utterances into the mouths of Pythagoreans.

First, as to the all-generating "four." In the beginning, according to Empedocles, was chaos, from which was precipitated a mysterious "ether." This was followed by fire and earth. Motion of the inchoate mass then generated water and air. Fire acting by a divine alchemy on air crystallized out the celestial sphere of the fixed stars. The stars had already been ejected as sparks (or rings?) from the ether; the reaction with air studded them immovably in the outermost of the heavenly spheres. There they remained stuck till the eighteenth century, when (1742) Newton's versatile friend Halley by accurate observation discovered that certain stars have proper motions. This is but one instance of many in which a simple piece of man-made scientific apparatus, here the telescope, has obliterated a theory built on insufficient or misinterpreted evidence.

Empedocles gave further particulars about his elemental astrophysics, only one of which turned out to be a fair guess: the sun is a fire. He had no conception of how hellishly hot that fire might be. But then, neither had the astrophysicists of the nineteenth and early twentieth centuries. New ideas of physics had to be invented before a reasonable conjecture was possible.

Another of the philosopher's pronouncements may be recalled for its curious resemblance to a remarkable speculation of Einstein's relativistic cosmology. The stellar universe, according to Empedocles, is not suspended in an infinite void, but is circumscribed, at a very great distance, by a vast mass of inert matter. In one theory of the universe deduced from general relativity it is required to make sense of the assumed cosmological mathematics. This is accomplished by postulating a "mass horizon" at "infinity"—substantially what Empedocles imagined without resorting to mathematics at all. Pythagorean numerology teems with these startling anticipations of current speculations. Perhaps only a numerologist will see in them more than accidental historical puns. To bring the record up to date it should be noted that the mass horizon has been abandoned at infinity—where human observation cannot penetrate—by a majority of competent cosmologists.

The next is of special interest, as it hints how easily the Pythagoreans and many of their successors slipped from science to theology or from theology to science. Until recently a cosmology which did not include the deity simply was not science.

In accounting for his four elements Empedocles first imagined a swarm of infinitesimal atoms, all alike and spherical. Why spherical? Because Pythagoras had asserted that of all solids the sphere is the one perfect, just as the circle is the one perfect curve, and the deity sanctions only perfection in creation.

Here we note another of these curious historical puns. The "billiard ball" atom of Dalton (1766-1844) served chemistry faithfully and well till the early years of the twentieth century. The spherical atoms of Empedocles were endowed with love and hate or, as a good Daltonian would have said, with selective chemical affinity. (Atomic love and hate are strict numerological deductions from the relations $2 = 1 + 1$, $4 = 2 + 2$, from which any Pythagorean could easily have derived them. Possibly Empedocles obtained them in this manner. The proof may be left to the ingenuity of anyone interested. It should not be difficult after scanning some of the examples given in the next chapter.) If the atoms were all alike, how did they generate the four distinct elements? By motion. And what caused the motion? The Divine Fire, or the Eternal Mind.

The jostling of the atoms aroused their dormant emotions of love and hate in varying intensities, causing different numbers of atoms to cling together or to repel one another. The attractions and repulsions were so justly balanced that all the primordial atoms cohered in precisely four elements. If the reasoning here is obscure it may be clarified by the remark that in the Pythagorean theory of values the number four is justice.

In one phase of elemental existence love may be dominant, in another, hate. When love is ascendant the elements and the material things composed of them are stable and endure; when hate is the stronger, disintegration supervenes. The atoms themselves are indestructible and eternal—their existence in time had no beginning, nor will it have an end. All the infinite diversity of material things is but a manifestaton of love and hate, and therefore ultimately of the self-moving Divine Mind.

Even the human soul is included in the grand synthesis. It consists of two ($2 = 1 + 1$) sections; a sensory part, generated in the same way as the elements, and a reasoning part, the latter an emanation of the Soul of the Universe. The rational part of the soul is not free during the life of the body, but is shut up in its elemental prison to expiate the sins of its previous incarnations. An evil life may condemn this rational part to spend its next sojourn on earth in the body of an unclean beast, or even in a cankered tree or a noxious weed. At this point there is a sinister hint of eternal punishment, in the everlasting and immutable "law of necessity." It would seem that if a soul is predestined or "fated" never to achieve final purification and so to cleanse itself as a necessary preliminary to reabsorption in the Universal Soul, it can never hope to escape the Wheel of Birth.

To round out this sketch of the four elements we recall another of those fortuitous historical puns. In the 1870's the

Scotch physicist P. G. Tait (1831-1901) reasoned that as the atoms were all spherical and kinetically alike, therefore they could not be the offspring of chance, but must have been "manufactured"—it was an age of machine production of uniform commodities. Hence there must have been an intelligent manufacturer, and therefore a supreme reason supervising the creation and continued operation of the universe. The Daltonian billiard-ball atoms on which Tait reasoned in the Empedoclean manner proved inadequate for the physics of the twentieth century. They were abandoned when it became necessary to assume that atoms are neither spheres nor all alike.

This revolution in atomic physics did not imply that Tait's theology was incorrect. It merely exemplified the historical facts that wisdom may be reached by many paths, and that the science and fable of one epoch may exchange places in another.

As a parting tribute of respect to a majestic Pythagorean, we recall that Empedocles fathered a comprehensive theory of organic evolution. According to him the plants evolved from the lifeless earth first. The animals then came up in segments, limbs here, heads there, to be united by the attractive force of love. Naturally a large number of monstrosities were produced in the process, but happily few of these survived. Men and women followed next, being ejected at the surface of the earth as unformed lumps or clods of matter by the pressure of the fire under the earth. The lumps congealed into the various members of the body and were assembled as in the creation of animals. As it is not yet known how life reached its present state, we need not be too severe on the myth which satisfied Empedocles and, with unimportant modifications, his Pythagorean successors. It was a consistent attempt to give a rational explanation of the origin of living things; and if anyone has yet done more, he has not published the outcome of his researches. In passing it may be mentioned that complete expositions, with colored pictures, of the Empedoclean theory of evolution are still frequently given in circus sideshows to evidently appreciative audiences. Not all of Empedocles' biology was as fanciful as the specimens exhibited. He is credited with some acute observations in physiology. But as these are in no sense mathematical, or even numerological, we pass them.

There is one detail, however, which is of capital importance for the sequel, especially in connection with Plato's high estimate of the value of training in mathematical reasoning as a preparation for apprehending eternal truths. Only reason—the better half of the soul—can reveal the truth of anything. The sensory half is fallible, deluding the reason through illusions, and is not to be trusted. In another guise this amounts to the reliance on pure reason rather than observation and experiment in science, the creed of all Pythagoreans both ancient and modern.

CHAPTER - - - - - - - - - - - - - - 14

The Cosmos as Number

INDIFFERENT to the rapidly gathering storm of popular resentment against their secrecy and exclusiveness, the Pythagoreans continued to elaborate their solution of the universe during their last months in Croton. Ignoring Cylon and his democratic insurgents, the Brotherhood proceeded with their unworldly theory as if they had all eternity instead of a matter of weeks in which to complete their task. Perhaps they were wise. What they succeeded in finishing and transmitting to their intellectual posterity was to prove less suggestive than the disordered mass of fanciful speculations they left undeveloped. Minds akin to their own still seek inspiration in the unfinished business of the Pythagoreans. Others may regret that Cylon and his mob could not reach all the Pythagorean colonies to exterminate them too as they exterminated the parent organization in Croton. More objective critics, even while unsympathetic to the master's teachings, merely recount what the Brotherhood propagated in the name of reason and let the facts be their own commentary. Of these unbiased judges Aristotle (384-322 B.C.), only about two and a half centuries later than Pythagoras himself, is the most explicit.

"Their training being exclusively in mathematics, which they were the first to develop, the Pythagoreans imagined the principles of mathematics to be those of everything."

This in itself would seem to be sufficiently devastating. But Aristotle, not being a devotee of mathematics himself and having almost no feeling for the subject, thought it necessary to present some particulars.

"Since naturally numbers are prior to everything, the Pythagoreans imagined they perceived closer analogies between numbers and things than between fire, earth, or water [three of Empedocles' four elements] and things. Thus justice was one combination of numbers, intelligence and reason was another, opportunity yet another, and so on.

"Again, they observed that the properties . . . of musical scales are expressible in terms of numbers. Because everything else appeared to have the forms of numbers, and because numbers in nature seemed to be antecedent to things, they concluded therefore that the elements of numbers are identical with the elements of things, and that the heavens are a number and a harmony.

"Having pointed out the close analogies between numbers and astronomical phenomena, and indeed between numbers and all phenomena of the entire cosmos, they constructed a system of astronomy. If any gap appeared in the system, they did their utmost to restore the connection [between numbers and the observable facts of astronomy]. For example, since ten seemed to them to be the number of perfection, they asserted that there are ten heavenly bodies [including the sphere of the "fixed" stars]. Only nine being visible, they imagined a tenth, the Counter Earth, to balance the Earth. . . . They maintained that number is the origin of things, and the cause both of their material existence and their modifications and different states. . . ."

This somewhat caustic summary of the Pythagoreans' comprehensive solution of the universe is so rich in half-hidden allusions that it must be taken apart piecemeal to disclose its damning finality. To anticipate, the numerological thread tying all the disparate items mentioned into a compact unity is the indisputable fact of arithmetic that $10 = 1 + 2 + 3 + 4$. Each of the numbers, 1, 2, 3, 4, 10 in this fundamental relation of Pythagorean numerology has not merely one meaning, or even two, but literally dozens of meanings, no pair of which has anything in common.

If this seems too nonsensical for the foundation of a rational system of the universe, an analogy with modern physical science may soften the harshness of any hasty condemnation we might be moved to pass on the science of our predecessors. Scanning each of several advanced treatises on the various divisions of classical physics—mechanics, heat, sound, light, electricity and magnetism—we note that two or more of them contain at least one pair of equations identically the same except possibly for the letters in which they are written. Now if a particular equation appears, say, in both the theory of electromagnetism and the theory of elasticity, we might describe certain phenomena of electromagnetism in the language of elasticity with which we may be more familiar. Or if the equations summarizing the vibrations of an elastic solid appear also in the theory of light, we may describe light as a vibration of a hypothetical elastic medium and call this medium the universal ether. Proceeding, we may even persuade ourselves that this ether has as factual an existence as a tangible lump of cobblers' wax. All this is substantially what the nineteenth century physicists did, even to the lump of wax. Their successors in the twentieth century abandoned the ether when Einstein and others convinced them that it does not exist.

The Pythagoreans—including modern believers in the ether —reasoned from the existence of mathematical analogies to the reality of mysterious entities behind the analogies. The ancients used nothing more advanced than elementary arithmetic and the simplest geometry in constructing their analogies. The moderns have used all the intricate machinery of mathematical analysis developed by the master mathematicians from Newton to the present.

Though the vocabulary, the grammar, and the syntax have changed, the thought embodied in the language has remained the same. And if the Pythagoreans read into the language of elementary arithmetic more than it may meaningly express, some of the modern equaled or surpassed them in their more recondite interpretations of a mathematical language descended from that arithmetic. The thought animating both readings was the dream that mathematics of itself can reveal the constitution of the universe and disclose the laws of nature.

To take a specific instance of the main point at issue, we may consider the status of Antichthon—the hypothetical Counter Earth dismissed by Aristotle with the contempt which no doubt it merited even in his day. The important thing is not this or that particular article, right or wrong, of the Pythagorean scientific creed. All such details lost whatever significance they may have had for science many centuries ago, and their only interest today is as curiosities or pathological excrescences of rational human thought. In themselves they are trivial. But the faith which gave birth to them is neither trivial nor outdated. It is livelier and more prolific of new knowledge than it ever was in the past and, as in the time of Pythagoras, it continues to prophesy verifiable fact and unverifiable fable. That faith is simply the belief that it is possible in some slight degree to predict the knowable and to foresee the future of the material universe. Ancient magic claimed to be able to do these things but never did. Less ancient astronomy had a considerable success. Modern science has had more successes than failures in its most highly developed departments, notably in physics, astronomy, and genetics; and in both the successes and the failures mathematical reasoning has played an impressive part.

Occasionally, as in relativity and the modern quantum theory, successful predictions have surprised even the men making them. An older success was the discovery of the planet Neptune in 1846 consequent on a mathematical analysis of irregularities in the orbit of Uranus. The mathematicians told the astronomers where to look for the new planet and it was found. This was a triumph for the mathematics of the Newtonian theory of gravitation. The numerology of the Pythagorean theory of the solar system predicted the existence of Antichthon, which of course was not observed in the heavens and never will be. But the faith inspiring the prediction was the same as that which led to the discovery of Neptune. As a recent (1918) instance of the same faith inspiring a prediction as false as that of Antichthon, a beautifully reasonable modification of general relativity predicted that the atoms of the chemical elements should exhibit certain characteristics. The observed fact that no such characteristics exist put them in the same category as Antichthon.

With these parallels between the science of the past and that of the present as a softener for the hard mysteries to follow, I shall present a few fragments of the "everything" in the Pythagorean "Everything is number." If any scientist of today expects or hopes for any sympathy from his successors of only a century hence, he will not be too contemptuous of this earliest attempt, twenty-five centuries ago, to give a rational account of the cosmos, but will grant it the courtesy of an amused tolerance.

The heart and brain of the Pythagorean cosmos are the decad and the tetrad. The decad consists of the first ten natural numbers, 1, 2, 3, 4, 5, 6, 7, 8, 9, 10, and the tetrad of the first four, 1, 2, 3, 4.

It may be emphasized at the outset that 1 sometimes was not accorded the dignity of being a number at all. But when some tremendous generalization required 1 to be a number to avoid irritating contradictions, 1 temporarily became as much of a number as the rest.

Though this ambivalence deprived 1 of some of its numerical privileges, the defect was more than compensated by the ascription of powers not shared by any other number. For obviously 1 is the author and progenitor of both the tetrad and the decad: $2 = 1 + 1$, $3 = 2 + 1 = 1 + 1 + 1$, and so on. Thus 1 may be identified with the universal and omnipotent One, the Creator of all things, when it shall have been shown that everything in the universe is generated by, or is implicit in, the decad. The cogency of this logic must be admitted.

Actually it will be sufficient to get as much as may be desired out of the tetrad, since the tetrad generates or begets the decad: $2 = 1 + 1$; $3 = 1 + 2$; $4 = 2 \times 2$; $5 = 2 + 3$; $6 = 2 \times 3$; $7 = 3 + 4$; $8 = 2 \times 4 = 2 \times 2 \times 2$; $9 = 3 \times 3$; $10 = 1 + 2 + 3 + 4$. Only a few of the many possible generations of the decad from the tetrad have been exhibited. Those chosen were among the most suggestive to the Pythagoreans. Others equally potent were $5 = 2 + 1 + 2$, $7 = 3 + 1 + 3$, $9 = 4 + 1 + 4$, where the common characteristic is apparent.

It should be noticed that no even number can be similarly decomposed into a sum of three numbers, of which the middle one is 1 and the first and last are the same. Trivial? Not at all. This truism of elementary arithmetic will appear as the numerological essence of the metaphysics of the Limited and the Unlimited, of the Finite and the Infinite, of Time and Eternity, which certainly are among the topics most frequently debated by metaphysicians all down the past two thousand years. If everything is number need anyone be shocked or astonished that metaphysics is a kind of mystical arithmetic?

To the uninitiated it may seem rather strange that numbers greater than ten are loftily ignored. But really they are not. For, as Pythagoras observed, "the decad contains all things; since numbers beyond the decad merely repeat the first ten." The thought there seems to be that $11 = 10 + 1$, $12 = 10 + 2$ . . ., $19 = 10 + 9$, $20 = 2 \times 10$, $21 \doteq 2 \times 10 + 1$, . . ., $29 = 2 \times 10 + 9$, . . . and so on. A Babylonian numerologist would have made all the numbers beyond 60 echo the truths implicit in the numbers 1 to 60. What the Pythagoreans did amounts to a special case of a device used in the modern higher arithmetic. They separated all the natural numbers into ten classes. The first class contains all the natural numbers which leave the remainder 1 when divided by 10; the second class contains all those which leave the remainder 2 when divided by 10, and so on, till the tenth class, which contains all the natural numbers that are exactly divisible by 10. For the purposes of numerology it was unnecessary to discriminate between the numbers in any one of the ten classes because all of them, by hypothesis, were numerologically indistinguishable.

The next fundamental assumption of the Pythagoreans lies much deeper, so deep in fact that civilized man can scarcely hope to fetch it up to the full light of reason. Odd numbers are male; even numbers, female. We can only ask why, expecting no answer except possibly a hesitant allusion to a vestigial phallicism or a forgotten Orphism. Primitive peoples seem to be even more solicitous than some of the moderns about sex, frequently incorporating it bodily and spiritually into their religions. Possibly the male 1 and the female 2 were sacred relics of some forgotten creed. Whatever may have been the origin of this physiological arithmetic, it is indispensable in the Pythagorean theory of the universe.

From the postulate that numbers are of opposite sexes it followed—for the Pythagoreans—that the male marriage number is 5 and the female marriage number 6, both of which fall back as they should in the all-containing decad. The reasoning here is simple. In lawful marriage one female is united with one male. But 2 is the first female number, and 3 the first unequivocally male number. This is one of those numerous occasions where 1, though not even and therefore presumably male, is denied a privilege granted the other numbers. The union of 2 and 3 is $2 + 3$, or 5, which accounts for the male marriage number. Its female companion is equally reasonable. For in marriage a female is multiplied by a male: $2 \times 3 = 6$.

If it be asked why $3 + 4$, or 7, is not the male marriage

number instead of $2 + 3$, Pythagoras replies that 4 is justice, and justice is clearly a male virtue, not female as required to produce a marriage with the male 3. Pressing him a little, we ask why 4 is justice. This is easily answered: $4 = 2 \times 2 = 2 + 2$, where it is to be ignored for the moment that 2 is female. But whatever the sex of 2, either of 2×2, $2 + 2$ expresses "the return of like for like" or, in more concrete symbols, "an eye for an eye and a tooth for a tooth," one of the immutable canons of all savage justice. Also, as will appear elsewhere, 7 is virgin and therefore inappropriate as the male marriage number.

We are in a dream world where anything we wish to prove can be proved, for the sufficient reason that any obstacle to strict deduction may be abolished by introducing as a new postulate the nonexistence of the obstacle. Our creative powers are unlimited. Nothing can balk us, because never once do we subject our conclusions to the drastic test of reproducible experience in the sensory world. Indeed they are beyond any such test. Our dream, a creation of the free reason, is strictly rational whether or not it has any counterpart in the everyday world of the senses. If now we follow Pythagoras and Empedocles and postulate that only the reasoning part of the soul can reveal the truth to mankind, we must believe with Plato that our dream world is the real one and the other an illusion.

And if we are inclined to censure the Pythagoreans for tampering with their postulates whenever they encountered a difficulty in their deductive numerology, we may remember that a similar practice is by no means uncommon in modern science. To take a simple and frequent instance, an ambitious mathematician attacks an outstanding problem. The solution of his problem would be a significant advance in science. But after several months of barren labor he finds that the real problem is beyond his powers. So he returns to the very beginning, makes a barely noticeable change in one of the given conditions of the problem which has blocked him at every step, and proceeds without difficulty. He then tries to persuade himself that the easy problem he has solved is as significant for science as the hard one he abandoned. As it would be invidious to cite current examples, we return to the first great masters of the art of replacing the difficult by the easy.

To round out the Pythagorean numerology of marriage we must account for the children. Now almost anyone manipulating numbers will chance upon a most remarkable property of 6—the female marriage number—even as the Pythagoreans happened on it early in their career: $6 = 1 + 2 + 3$. But 1, 2, 3 are all the numbers less than 6 that divide 6 without remainder. That is, 6 is the sum of all its divisors less than itself. For this reason the Pythagoreans called 6 a perfect number. Such numbers are extremely rare and hard to find, and it is not known even today whether such a thing as an odd perfect number exists. The next perfect number after 6 is 28, since $28 = 1 + 2 + 4 + 7 + 14$, and 1, 2, 4, 7, 14 are all the divisors of 28 less than 28 itself; the next is 496; the next, 8128.

In the perfection of $6 = 1 + 2 + 3$ Pythagoras saw the temporarily male 1 uniting with the permanently female 2 and the ever-mystical 3 in perfect marriage. Why the 3? Because this 3 is the first and commonest of all the innumerable trinities that have dominated religions since the dawn of history, namely, the human trinity of Father, Mother, and Child—1, 2, 3. Nor is this all. The child is the union of its father and mother, $3 = 1 + 2$. All this no doubt is merely fantastic so long as we deny it any pathos. The number 3 symbolizes the eternal trinity; and it is rather touching that the early numerologists should have endowed a transitory human happiness with the permanence they wished it to have.

In passing we observe that 3, being the first fully male number, may be man. Man therefore is somehow divine, since 3 is also the holy trinity. This particular deduction is found in early Christian numerology. The Pythagoreans noticed an even more suggestive consequence of the identification of 3 with man. Nothing in human experience is more certain than the tragic fact that man's life has a beginning, a middle, and an end. But 3 is the only number in which beginning and end are evenly balanced against the middle, $3 = 1 + 1 + 1$. Thus man's fate to be born, to mature, and to die is implicit in the 3 which he is. Even Aristotle, for all his hard-headed superiority to the deluded Pythagoreans, was lured into numerology in his *Poetics*—and elsewhere. His demand that a tragedy shall have a beginning, a middle, and an end is unadulterated numerology. The master himself said it.

The perfection of 28 is even richer than that of 6 in cosmic truths, but we must pass them with a mere allusion. A week is 7 days; 14 is therefore 2 weeks and 28 is a lunar month; 1 and 2 are man and woman or God and woman; 4 is justice; and 7, the "virgin number"—so designated because 7 does not generate numbers within the decad either by multiplication or by division—is the union of 1, 2, and 4. And so forth. Nearly "everything," from man to the moon, is justly (fourly) one perfection.

In his summary of the ancient numerology Aristotle remarked that "justice was one combination of numbers, intelligence and reason another, opportunity yet another, and so on." We have seen how 4 was justice. It will be interesting now to glance at opinion and knowledge. Though not specifically mentioned by Aristotle, these are closely akin to the abstractions in his list. Much of what Plato and Socrates (speaking what Plato put into his mouth) had to say on opinion and knowledge was appropriated from the pioneer numerologists.

We enter this labyrinth of philosophical arithmetic through the double gateway of the Limited and the Unlimited—those mystical abstractions which were to be the alpha and omega of metaphysics from Plato to Hegel, and of mathematics from Pythagoras to Cantor (1845-1918), the founder of the modern theory of the mathematical infinite.

Odd numbers in Pythagorean numerology are limited, finite, and determinate; even numbers share in none of these masculine qualities of decisiveness. The meanings of the technical terms here differ from those current today. Thus "finite" means bounded, terminated, while "infinite" means unbounded, not ended. Both "finite" and "infinite" occur in modern mathematics with these definitions, but the connotations are not those of Pythagorean numerology or anything even faintly resembling them.

In the Pythagorean attempt at a rational science the "finiteness" of the odd numbers and the "infiniteness" of the even numbers signified two elementary facts about numbers which to us are trivialities. The odd number 5, for example, can be separated into two equal numbers and a unit, and the unit may be imagined in the middle of the split: $5 = 2 + 1 + 2$. Similarly $7 = 3 + 1 + 3$, and the general odd number is $n + 1 + n$. The creative One, 1, "bounds" or "limits" the two equal numbers. Similar separation of the male numbers by the female 2 is impossible, since an odd number when divided by 2 leaves a remainder (½) which is not a whole number. Numerologically, then, a female can separate two males but never one.

An even number, on the contrary, is not limited in its internal constitution by the divinely creative One. $4 = 2 + 2$, $6 = 3 + 3, \ldots, 2n = n + n$. Female numbers can thus be split by the least of them (2) into 2 whole numbers. A probable underlying reason for this use of "limited" and "unlimited" will appear in the next section, where it is seen that a "line" is "limited" by its ends, which are points, and a point is 1.

From all this it follows that the limited odd numbers are appropriate to constancy and knowledge, while the unlimited even numbers can express themselves only through inconstant opinion. It is best not to enquire into the details of the proof.

Not all of the numerology of the Limited is so fanciful as the preceding specimen. If everything is number as Pythagoras asserted, it must be possible to prove that all space is number. The Pythagoreans accomplished this by a most ingenious application of their theory of limitation. Their solution of the problem of space was the earliest attempt to give a consistent account of dimensionality. What does it mean to say that a certain space has one dimension, or two, or three? A satisfactory answer, valid for a space of any (finite or infinite) number of dimensions, was given only in the 1920's. Though the Pythagorean solution of the space riddle long ago ceased to make sense to mathematicians, Pythagoras and his disciples should be given some credit for having imagined a genuine problem. Without distorting the meaning of the word too violently, an impartial critic could say that even though their solution was wrong, it was rational. The solution was an important step toward the identification of the four material elements with numbers and geometrical figures. We proceed to the numerological proof that space is number.

Points are the primary elements of space for Pythagoras, and a point is that which has position only. Unlike material things a point has neither parts nor magnitude. These defects are shared by 1 when the latter is regarded as the Monad or the generative element of number. If Pythagoras thought of space as being made up of points, then points generated his space. But whatever he imagined space to be, he identified a point with 1.

A straight line, or briefly a line, in our geometry extends indefinitely in either of the directions determined by the line. But in Greek geometry a line was merely a finite segment of our line, and it was postulated that a line could be extended to any desired (finite) length. Thus a Greek line had two ends, each of which was a point, or 1. So in Pythagorean numerology a line is 2. We see also why an odd number is "finite" or "limited." For example, in $7 = 3 + 1 + 3$ the 1 is the point limiting the 3's.

However space may be defined, it is advantageous to abstract a part of the definition from the intuitive notion of extension as on a flat surface. The Pythagoreans had defined a line as a length without breadth—possibly an innovation of Thales. So neither the point 1 nor the line 2 was "space" for a Pythagorean; and it is illuminating to observe that neither the 1 nor the 2 enjoys all the privileges of the lordly male (odd) numbers. But with the fully masculine 3 we reach the real, limited numbers and therefore also (we hope) plane space. Actually we do, because precisely 3 points not on one line are necessary and sufficient to determine any particular plane. Indeed it is sufficient that an equilateral triangle be designated in order to specify plane extension, and such a triangle is fixed when its three vertices are assigned. Each vertex is a point, or 1. The triangle is the union of its three vertices, $1 + 1 + 1$, which is 3. Thus a plane is 3.

Suppose, however, that we had counted the sides of the equilateral triangle instead of its vertices. Each of the sides is a line and therefore is 2. As there are 3 sides it would seem that the triangle is 3×2, or the perfect 6. But this is not so. The fallacy is that each end of any one of the sides is counted twice, once on each of the lines terminating at that end. So we must divide 6 by 2. The outcome is again 3 as the number of the plane. This check on the correctness of the logic must have given Pythagoras a moment of ecstasy.

At the next stage a grand new principle emerges in the numerology of space. A line, we have seen, is bounded or "limited" by points. The primary element (the point, 1) of all space thus appears as the limiting element of the secondary element (the line, 2). This suggests that the secondary element should appear as the bounding or limiting element of the tertiary element of space, namely, the triangle. It does: the triangle is bounded by 3 lines. With this enticing hint of a stupendous general law immanent in all space, Pythagoras anticipated demonstration and boldly conjectured that solid space, the space of material bodies, is the number 4. Then, like a conscientious scientist, he tested his guess against what appeared to him to be the facts. If they confirmed him, he would be the happiest man on earth.

The simplest of all regular solids is the tetrahedron, which has 4 points as its vertices and 4 equilateral triangles as its faces. The grand principle of limitation by elements of the next lower order is sustained. But there is more, much more. Solid space, we have just proved, is 4, and 4, being justice itself, casts no doubtful shadow. The 4 triangular subspaces bounding and limiting the tetrahedron are themselves bounded and limited by the 6 lines which are the edges of the solid, and the number 6 is perfect. Moreover the 4 vertices of the tetrahedron limit the 6 lines limiting the 4 triangles limiting the solid. Being thus limited in all conceivable ways the tetrahedron, and therefore also solid space, is essentially male in spite of all its 2's, its 4's, and its perfect 6.

Everything is now accounted for: 1 is the point, 2 the line, 3 the plane, and 4 the solid. But what unforeseen miracle is this? The 1, 2, 3, 4 are the tetrad; their union, that is, all space, is the number of the decad itself: $1 + 2 + 3 + 4 = 10$. Since all material things exist only in space, they too are number, and the tetrad generates them all. Pythagoras was the happiest man on earth.

Continuing with the master's "everything," we shall pass on in a moment to the "things" mentioned in Aristotle's indictment—"they concluded therefore that the elements of numbers are identical with the elements of things." The immediate problem is to derive matter from the tetrad 1, 2, 3, 4. Possibly the most appreciative exposition of the relevant numerology is to be found in Plato's dialogues. Without implying that Plato himself took all of this Pythagorean physics and chemistry as seriously as he might wish to have us believe, we may note in passing where he probably got it.

About the middle of the fifth century B.C. the erudite Pythagorean scholar and philosopher Philolaus (flourished *c.*

450 B.C.) compiled a comprehensive summary of the master's teachings. At that time the Brotherhood had been disbanded for about half a century. As we shall see in the next chapter the dissolution of the Pythagoreans as an organized secret society was the outcome of Cylon's revenge. But though the Brotherhood had ceased to exist as an active political body, some of the brothers who had known the master in the flesh were still living in the intellectual colonies established originally by the parent organization in Croton. These ageing survivors of one political purge after another were in somewhat the same situation as the intellectual Jews in Europe during the Nazi régime.

Suspected of all sorts of mischief of which they were innocent, and charged with crimes against the ruling tyrannies they had no intention of fighting in their desperate circumstances, the harassed Pythagoreans resorted to stealth to keep their science alive though they themselves might perish. The affected secrecy of their prosperous years became a practical necessity if their teachings were not to die with the Brotherhood. Consequently but few written expositions of the Pythagorean science and philosophy were compiled, and these few passed from hand to hand only under the most solemn pledge of secrecy. The summary of Philolaus is said to have been the fullest and most accurate of all. Even in Plato's early years, when active hostility to the Pythagorean sect was a thing long past, the Pythagorean "bible" of Philolaus was extremely difficult to procure. Plato is reputed to have obtained his copy from Archytas of Tarentum. Archytas himself was an enthusiastic scholar of Pythagoreanism. Recognizing a kindred mind in the young and impressionable Plato, Archytas generously presented him with his priceless copy of the Pythagorean bible. (One account says Plato paid a high price for the book, but for several reasons this is unlikely.) Except for a few fragments of very doubtful authenticity the work itself is no longer extant, but allusions to its contents occur in the writings of Greek historians.

The loss of Philolaus' compendium is compensated by the assiduous study Plato evidently made of its profundities, especially of its numerological science as presented for example in certain sections of his *Timaeus*. In selecting our samples of Pythagorean physics, chemistry, and astronomy, we have drawn occasionally on the riches in Plato's dialogues, to which anyone interested is referred for further particulars. Aristotle also has illuminating—if unsympathetic—comments on the science of the Pythagoreans. But by reasonably credible tradition the later Greek historians and philosophers based some of their accounts on the bible of Philolaus.

Written about fifty years after the death of Pythagoras, how reliable was this primary source? The question is similar to that regarding our own gospels, said by some critics to have been compiled not earlier than seventy or eighty years after the crucifixion. On the whole then it seems that we shall not go too far astray if we credit Pythagoras and his disciples with as much as Plato granted them.

By drawing on Plato's account of Pythagorean science here, ahead of its chronological order, we shall be in a position when we come to his own etherealization of Pythagorean numerology to see it "steadily and whole" by itself.

It is to be shown that all material things are numbers. In the proof (numerological, of course) that animals are numbers we catch a glimpse of prehistoric art. The number of any animal, or of any class of animals, such as "man" or "horse," is found by a uniform anatomical arithmetic. A diagram of a man, say, is outlined in the dust. Now man has certain distinguishing parts—two hands, two feet, one head, one heart, and so on. On the regions of the diagram corresponding to these parts pebbles are placed, one on each region. The total number of pebbles is the number required. Incidentally this is an instance of calculation in the original meaning of the word, for calculus is derived from the Latin for a pebble.

A recent (1942) observation by an English naturalist in India suggests an even earlier origin of the schematic representation of human anatomy. According to this observer the beginning of art was pre-human. It seems that the monkeys infesting a certain Indian village have appropriated a flat hilltop in the vicinity for their cultural activities—capering, courting, and the rest. Every so often one of the sportive monkeys will suddenly interrupt his dance, squat down hurriedly, press his open left hand firmly into the dust, and with a stick held in his right hand as a draftsman holds a pencil, rapidly trace a line round the impress of the left hand. Then, evidently somewhat fearful that he has committed an unnatural sin, the artist leaps to his legs and skips away to the nearest tree. The other monkeys then prance round the masterpiece, viewing it with apprehensive fascination. Is it a real hand, or is it the abstraction of all hands, the universal Hand in the realm of Eternal Ideas? Like ourselves they cannot make up their minds. They resume their accustomed occupations.

Intermediate between the pebble-calculus of living things and the more abstruse semigeometrical numerology of the four elements, is another Pythagorean calculus of which a considerable part has lasted into the unmystical higher arithmetic of today. It is illustrated in the story of the merchant whom Pythagoras asked if he could count. On the merchant's replying that he could, Pythagoras told him to go ahead. "One, two, three, four . . . ," he began, when Pythagoras shouted, "Stop! What you name four is really what you would call ten. The fourth number is not four, but the decad, our tetractys and the inviolable oath by which we swear." To have satisfied Pythagoras the merchant should have counted (in our numerals) 1, 3, 6, 10, 15, 21, 28, 36, . . . These are the triangular numbers, so called because when represented as pebble-patterns they are equilateral triangles.

Pythagoras constructed these numbers thus:

```
                                   .
                    .            .   .
        .         .   .        .   .   .
 .    .   .     .   .   .    .   .   .   .
```

and so on. The next, 15, is formed by bordering the 10 triangle along any one of its sides with 5 more pebbles; the next, 21, by similarly disposing 6 pebbles, the next adds 7 more pebbles, the next 8, the next 9, and so on.

Square numbers are pebble patterns constructed by the obvious steps in

```
                          .  .  .  .
               .  .  .    .  .  .  .
       .  .    .  .  .    .  .  .  .
 .     .  .    .  .  .    .  .  .  .
```

where 9 is obtained from 4 by bordering on any two adjacent sides of the 4, and 16 is the bordered 9. The next, 25, is the bordered 16, and so on forever. In the same way any other regular plane figure (all sides equal and all angles equal) provides the basic framework for a pebble-pattern of a class of so-called polygonal numbers—pentagonal, hexagonal, heptagonal, octagonal, and so on for as long as we please.

This connection between regular geometrical figures and the corresponding sequences of numbers was profoundly significant to the Pythagoreans and after them to the Platonists, partly because of the evident union of space-symmetries with numbers, partly because the tetrad and the decad kept turning up unexpectedly in various disguises. There were also oblong numbers corresponding to patterns of pebbles arranged in rectangles with sides differing by 1 pebble, for example $30 = 5 \times 6$. When Pythagoras observed that an oblong is equal to twice a triangular number, as in $30 = 2 \times 15$, his enthusiasm was unlimited.

Encouraged by his spectacular successes with plane figures, Pythagoras boldly ventured into solid space. There, in imagination, he pebbled out the successive cubic numbers 1, 8, 27, 64, 125, . . . by a uniform process which may be left to the ingenuity of the reader to rediscover. And there he stopped, because space for him as for all the Greek numerologists and geometers had only three dimensions. They could visualize the result of multiplying three numbers together as the volume of a solid. Thus $3 \times 4 \times 10 = 120$, is the solid content of a box whose edges are 3, 4, 10. But a multiplication such as $3 \times 4 \times 10 \times 12$ baffled them in their geometrical arithmetic, for to "multiply four lines" had no meaning in their three-dimensional space. All such artificial barriers vanished into nothing when algebra supplanted geometry as the language of number. But the triangular and other polygonal numbers of the Pythagoreans, also the cubes, have survived at least as names in the modern theory of numbers. The oblongs dropped out of the vocabulary long ago.

The most significant detail for Pythagorean science in all this is the fourth of the triangular numbers, 1, 3, 6, 10, 15, 21. . . . It is 10, or the decad. But it is also a triangle, and therefore is the sacred tetractys. Since all things are contained in the decad—according to Pythagoras—we see perhaps why ten was the perfection of all numbers and, according to Plato, the archetypal pattern of the universe. We see also a preliminary hint of the Platonic generalization that matter is composed of triangles. This will be confirmed when the 4 elements are generated from the 4th member of the sequence of triangular numbers, namely, from the triangular decad. It is not surprising that the Pythagorean Brotherhood made 10 —really the 4th triangle—their oath and their most jealously guarded secret. He who swore by the tetractys and betrayed his oath was damned indeed, for he had betrayed the entire universe of which he himself was necessarily a fraction—a Greek would have said a ratio.

Though it would be amusing to unravel all the tangled numerology in Plato's account (particularly in the *Timaeus*) of the creation and structure of the material universe, it is not necessary to do so in order to obtain a sufficient idea of Pythagorean chemistry, physics, and cosmogony. Perhaps enough has already been given to suggest the possibilities in such a typical passage as the following.

"Now that which is created is of necessity corporeal and also visible and tangible. And nothing is visible where there is no fire, or tangible which is not solid without earth. Therefore the deity in the beginning of creation made the body of the universe to consist of fire and earth. But two things cannot be held together without a third; they must have some bond of union. Now the most beautiful bond is that which most completely fuses into the things bound. Proportion is best adapted to effect such a fusion. For whenever among three numbers, whether solids or any other dimension, there is a mean, so that the mean is to the last term as the first term is to the mean, and when (therefore) the mean is to the first term as the last term is to the mean, then, the mean becoming both first and last, and the first and last both becoming means, all things will of necessity come to be the same, and being the same, all will be one."

Undoubtedly this is a transcript from the lost Pythagorean bible of Philolaus, for it is the purest of the pure Pythagoreanism. To understand what is meant, it is helpful to translate the rather involved language into its simple equivalent in terms of elementary arithmetic. Actually the passage refers to certain obvious properties of common fractions. Rather, the arithmetic involved is obvious to us. But it was not obvious to a Pythagorean of the fifth century B.C., or even to a Greek mathematician of Plato's time, neither of whom had a readily intelligible way of writing fractions. Curiously enough this somewhat obscure passage would have been plainer to a schoolboy of the eighteenth century than it probably is to college graduates today.

Except in old-fashioned textbooks we seldom meet with "ratios" and "proportions" in modern scientific writing. The "ratio" of the number m to the number n is written $m:n$, and is merely the fraction which we write m/n or $\frac{m}{n}$. If the ratio $m:n$ is equal to the ratio $r:s$, the antiquated way of writing this is $m:n::r:s$; the modern way $\frac{m}{n} = \frac{r}{s}$ or $m/n = r/s$. Even the old way is easier to understand than what the Pythagoreans and their Greek successors used. They had no expressive mathematical symbolism like ours, but wrote out everything in words, as in the above extract from Plato. The relevant details here are his "proportion" and "mean."

Four numbers, say, m, n, r, s, are "in proportion" when the first is to the second as the third is to the fourth or, in the language of ratios, when the ratio $m:n$ is the same as the ratio $r:s$. Thus m, n, r, s are "in proportion" if $m:n::r:s$ or, in our simpler notation, if $\frac{m}{n} = \frac{r}{s}$. In this "proportion" the outside numbers m, s, are the "extremes," the inside numbers n, r, the "means." The numbers m, n, r, s are the "terms" of the "proportion."

Many special cases arise. That in which the means n, r are the same, so that $r = n$ and $m:n::n:s$, was of great importance for the Pythagoreans, also for the Greek geometers. In this case n is called "the geometric mean" between the extremes n, s, or "the mean proportional" of n, s. Turning all this into the equivalent statements in fractions, we have $\frac{m}{n} = \frac{n}{s}$, and therefore as any grade-school pupil knows ("clearing of fractions") $m \times s = n \times n$ or, in elementary algebra, $ms = n^2$, and therefore the "geometric mean" (n) of two numbers (m, s) is the square root ($\sqrt{ms}$) of their product (ms).

The arithmetic in the passage from Plato is simply this. From the "proportion" $\frac{m}{n} = \frac{n}{s}$ it follows immediately that $\frac{n}{m} = \frac{s}{n}$ (if two fractions are equal, the results of dividing 1 by each of them are also equal). But this is what he is saying: from $m:n::n:s$ it follows that $n:m::s:n$, in which the "mean," n, of the original "proportion" has become both the first and the last in the second, and the first and last, m, s in the original proportion have become the "means" in the "proportion" implied by the original. So Plato's elementary arithmetic is correct.

To assist in the recognition of other disguised arithmetic in Platonic philosophy, the following definitions may be recalled. In the sequence of numbers.

$$1, 5, 9, 13, 17, 21, \ldots,$$

the step from each number to the next is the same, namely 4. The numbers are said to form an "arithmetic progression" with first term 1 and "common difference" 4. The arithmetic progression with first term 6 and common difference 5 is

$$6, 11, 16, 21, 26, 31, \ldots$$

It will be noticed that $16 = \frac{1}{2}(11 + 21)$, $21 = \frac{1}{2}(16 + 26)$, and so on; each number after the first is half the sum of its left and right neighbors. For this reason each number after the first is called the "arithmetic mean" of its immediate predecessor and immediate successor.

Now suppose we divide 1 by each number in a given arithmetic progression, say the second above:

$$\frac{1}{6}, \frac{1}{11}, \frac{1}{16}, \frac{1}{21}, \frac{1}{26}, \frac{1}{31}, \ldots$$

The resulting sequence of numbers is called a "harmonic progression," and each number after the first is said to be the "harmonic mean" if its immediate neighbors. As an example, one of the sequences

$$\tfrac{3}{4}, 1, \tfrac{5}{4}, \tfrac{3}{2}, \tfrac{7}{4}, 2, \tfrac{9}{4}, \ldots$$
$$\tfrac{4}{3}, 1, \tfrac{4}{5}, \tfrac{2}{3}, \tfrac{4}{7}, \tfrac{1}{2}, \tfrac{4}{9}, \ldots$$

is an arithmetic progression, the other a harmonic progression.

The third and last kind of progression repeatedly used by Plato is the "geometric," in which each number after the first is obtained by multiplying the preceding number by the same multiplier. Thus,

$$3, 6, 12, 24, 48, 96, 192, \ldots$$

is a geometric progression with first term 3, and multiplier, or "common ratio," 2. The "geometric mean" of 6 and 24 is the number, 12, between them; the geometric mean of 48 and 192 is 96, and so on.

A little algebra will show that if A, H, G are, respectively, the arithmetic, harmonic, and geometric mean of the numbers M, N, then

$$A = \tfrac{1}{2}(M + N), H = 2MN/(M + N), G = \sqrt{MN}.$$

A little more will reveal the simple fact—which delighted and mystified the ancient numerologists, including Plato—that G is the geometric mean of A and H. On the whole it seems fortunate that the Greek philosophers knew no algebra.

Now although all of Plato's elementary—and disguised—algebra may be trite and trivial to a modern schoolboy, any mathematician must admire the pertinacious ingenuity which first reasoned it all out verbally without mathematical symbolism of any kind. The statement that $A/G = G/H$, or $A:G::G:H$, was called the "perfect proportion" by the Pythagoreans. It is said to have been brought from Babylon to Croton by Pythagoras. Any mathematician with imagination enough to divest himself for a moment of all his acquired technique, and to think himself back to the rhetorical arithmetic of the sixth century B.C., may agree with the Pythagorean Brothers that the "perfect proportion" was no invention of man, but the masterpiece of the Great Arithmetician of the Universe Himself.

The harmonic progressions and the harmonic means implicit in the supreme discovery originated in the numerical expression of the law of musical intervals which inspired Pythagoras to his "Everything is number." He and his disciples then sought music and harmony in the four elements of all material things and in the heavenly bodies. That they found what they sought is not remarkable when we remember that all harmony, all space, all matter, and all bodies, celestial and terrestrial, are in the decad which is the all-creator. As Plato expounds the theory, continuing after his proof that all is one—and therefore the deity:

"If now the frame of the universe had been created merely as a surface without depth, one mean would have sufficed to unite it to the other terms. But as the world must be solid, and solid bodies are always compacted not by one mean but by two, the deity placed water and air in the mean between fire and earth and, so far as possible, made them to have the same proportions—as fire is to air, so is air to water, and as air is to water, so is water to earth. Thus he constructed and united a visible and palpable heavens. Out of the four elements he so created the body of the universe in the perfect harmony of proportion [the perfect proportion]. Being thus endowed with the spirit of friendship, and being at unity with itself, the universe was indissoluble by any hand other than its maker's."

Without going into details for this particular application of the general principle, we note in passing that "the spirit of friendship" refers to a curious property of certain rare pairs of numbers discovered by the Pythagoreans. If each of the numbers m, n is equal to the sum of the divisors of the other, m, n is called an "amicable number pair," and m, n are "amicable" or "friendly" numbers. As in counting the divisors of perfect numbers, a number itself is not reckoned as a divisor. The smallest amicable numbers are 220 and 284. The Pythagoreans regarded this intimate union between amicable numbers as the very essence of friendship and the innermost soul of harmony.

Plato balances his account of the creation with an equally numerological passage in which it is shown that the deity fashioned the world in the likeness of the universal Animal "comprehending within itself all other animals." The proof that this World Animal does not suffer death and corruption as do other creatures, including man in his perishable body, is somewhat obscure even as numerology. The part quoted, however, is of greater scientific interest—science being understood of course in the Pythagorean sense. A suggestion or two will suffice to clarify the numerology. What follows is the simple core of the Pythagorean cosmogony and cosmology.

A regular solid is a body having as its faces regular polygons all of the same shape and size. Precisely five regular solids are possible and can be constructed in our three-dimensional (Euclidean) space. The first is the tetrahedron, having four equilateral triangles as its faces; the second is the cube, or hexahedron, having six squares as its faces; the third is the octahedron, having eight equilateral triangles as its faces; the fourth is the dodecahedron, having twelve regular pentagons as its faces, and the fifth is the icosahedron, having twenty equilateral triangles as its faces. The tetrahedron, the hexahedron, the octahedron and the icosahedron, but not the dodecahedron, were known to Pythagoras. Hence when the early Pythagoreans mistakenly supposed that only four regular solids are possible and can be constructed (as Plato showed very neatly for all five), they were inspired to prove that these four are identical with the four primal elements of all material bodies.

As a tentative guess, which they fitted snugly into their cosmic numerology, they postulated that the four elements fire, air, earth, and water are, respectively, the tetrahedron, the octahedron, the hexahedron (or cube), and the icosahedron. The faces of all but the hexahedron are triangles. This blemish was easily obliterated by splitting each face of the hexahedron into two triangles by a diagonal of the square. Thus, as Plato observes in the *Timaeus*, all matter is essentially triangles. The respective numbers of triangles constituting the elements are 4, 8, 12, 20. These then, if the numerology demands it, may be taken as the numbers which "are" the elements fire, air, earth, water, respectively.

The "essential triangularity" of matter, identifying it with the mystical 3, suggested that 4, 8, 12, 20 be accorded the place of highest dignity in the numerology of Pythagorean chemistry. But other possibilities were not thereby precluded, provided the resulting numbers could somehow be attached to the four regular solids known to the Pythagoreans when they somewhat prematurely announced their theory of the elements. Allowing ourselves this legitimate latitude we might easily verify Plato's numerology by which "the deity placed water and air in the mean between fire and earth."

But if the object is to prove the capital theorem that all matter is triangles, it can be done more elegantly and more convincingly (to an expert numerologist) through the following 6 proportions, in which *F*, A, *E*, W are, respectively, fire (tetrahedron, 4), air (octahedron, 8), earth (hexahedron, 12), water (icosahedron, 20):

$$F:A::3:6, \qquad A:E::10:15,$$
$$F:E::1:3, \qquad A:W::6:15,$$
$$F:W::3:15, \qquad E:W::6:10.$$

The numbers appearing in these mysterious proportions are 1, 3, 6, 10, 15. But these are the first 5 triangular numbers, and hence are themselves triangles. Since there are possible exactly 5 regular solids and exactly 6 essentially different proportions of the kind shown, and because all 6 have been exhibited, it follows that the creation is perfect, as Plato asserts, because the number 6 is perfect. This completes the proof, which is itself perfect because it consists of 6 proportions. So far as is known this beautiful demonstration was overlooked by the Pythagoreans and their successors.

Now what can have prompted the Pythagoreans in the first place to identify the four Empedoclean elements of matter with the four regular solids they knew? Numerous answers have been given and as many more are easily imagined. Regretfully passing the ancient solutions of this profound conundrum, we shall state one by a master astronomer, mathematician, and numerologist of a much later epoch. Johannes Kepler (1571-1630) is immortal in the history of astronomy for his famous three laws of planetary orbits. It was by attempting (successfully) to account for Kepler's laws that Newton was led to his own law of universal gravitation. The sequel is too familiar to need retelling here. But it is interesting that only about seventy years separated Kepler's laws from Newton's, and that the first of these major prophets of science was a confirmed numerologist while the second, who built on the secure foundations laid down by the first, was constitutionally incapable of believing anything whatever of numerology. In this respect numerology and theology are alike: it does not necessarily make any difference to a man's science what he believes or disbelieves about either. Some of the leading twentieth-century scientific numerologists are as distinguished in science as are their opponents who have only disrespect for all number mysticism.

Kepler's solution of the problem of the elements was ingenious and charming. Of all the regular solids the tetrahedron encloses the smallest volume for its surface, while the icosahedron encloses the largest. But these volume-surface phenomena are qualities of dryness and wetness, respectively. Since fire is the driest of the four elements and water the wettest, the tetrahedron is fire and the icosahedron is water. To assist the memory Kepler decorated his diagram of the tetrahedron with the drawing of a bonfire, and his icosahedron with a lobster and some fish. Obviously—according to Kepler—earth is the cube. For if any material thing on this Earth can sit more foresquarely on its four-square bottom than a cube, the Absolute himself does not know what it may be.

Kepler decorated his earthy cube with a carrot, a tree, and miscellaneous gardening implements. The numerological and physical contrary of the super-stable cube is the unstable octahedron. For if we hold this solid lightly by two of its opposite vertices between a forefinger and thumb, and give it a fillip, it spins like a teetotum. By this appeal to experiment it is proved that the octahedron is as unstable as air. Therefore air it is. This aerial solid was decorated most appropriately with clouds and birds on the wing. Last there was the dodecahedron, the villain of Pythagorean chemistry, with its twelve ugly pentagonal faces to be accounted for. It could not be one of the four elements, for these had already been identified. What on earth could it be? Obviously nothing. It must be—as Plato realized long before Kepler—a "celestial entity." But the signs of the Zodiac symbolize the entire heavens. Because the Zodiac has twelve signs and the dodecahedron twelve faces, therefore the dodecahedron is the universe. The diagram of this solid was embellished with drawings of the Sun, Moon, and stars.

The heavenly dodecahedron was almost as great a disaster for the Pythagoreans as the discord that the square root of two is not a rational number. Their proof that the elements are four of the regular solids was probably complete when this fifth regular solid—the dodecahedron—presented itself. It was a most unwelcome visitor. Hippasus, one of the Brothers, is said to have introduced the new solid to his colleagues, who received it somewhat coldly. One legend asserts that Hippasus was cast adrift in a boat without sails, oars, or rudder to perish for his presumption in claiming the shattering discovery as his own and not the master's. Another hints that Pythagoras was so disconcerted by the dodecahedron that he ordered the death of its discoverer, and had him sewed up in a sack and pitched over a cliff into the sea. (The chronology here is all awry.) But it was impossible to drown the dodecahedron and the Brothers settled down to make the best of it. To explain how they found to their delight that after all it fitted into their numerology of the universe as justly as a well-made key fits its lock, would demand an analysis of Plato's famous "Nuptial Number." But this is too involved a riddle to be taken apart here. It is enough to say that the dodecahedron, instead of destroying the Pythagorean numerology of the four elements, confirmed it gloriously in every detail and disclosed unimagined harmonies of transcendent beauty in the prolific tetrad, the tremendous tetractys, and the all-generating monad.

Plato's disposal of the awkward dodecahedron is as poetical as his "treatment" of the four "elemental" bodies. "The fifth solid," he states, "is used to embroider the heavens with constellations." The dodecahedron is therefore not an element of matter, but is the elusive "quintessence"—the "fifth essence"—of all the elements and indeed of the universe. For further details we must refer once more to the *Timaeus*.

To return for a moment to the Earth before leaving the elements forever, we note a detail of the Pythagorean proof that earth is the cube, as this has come down into our own everyday speech. The Earth is "four-square." Why? Because it has 4 cardinal points, and the line joining North and South intersects that joining East and West at right angles, again a four-square phenomenon, and justly so, for 4 is the eternal justice of the One.

Quitting the Earth and all its elements, we follow Pythagoras into the higher realm of the heavenly bodies, to catch with him a bar of that celestial harmony of "the music of the spheres" which solaced Kepler in his darkest hours of poverty, domestic tragedy, persecution, and twenty-two years of discouragement as he calculated, calculated, calculated to discover the laws of planetary orbits. Only a Pythagorean faith in a numerical harmony of the universe sustained him in his grinding drudgery and urged him through one disappointment to another, till at last he triumphed beyond his most ambitious hope. If numerology comforted Kepler it may be pardoned whatever pranks it has played on the wilfully credulous.

We have seen how the law of musical intervals inspired Pythagoras to his philosophy of number, and we must now seek with him the resultant music in the divine decad. To his amazed delight Pythagoras discovered that the tetrad—1, 2, 3, 4—alone contains the celestial consonances. For with the "octave" of a note is associated the ratio 2/1, with its "fifth" the ratio 3/2, and with its "fourth" the ratio 4/3. These basic empirical facts of harmony were discovered (probably) by adjusting the movable bridge of the monochord and plucking the varying segments of the string.

Though the argument is too involved for analysis here, the Pythagoreans, and after them Plato, deduced from this elementary acoustics that the universe has a soul and that the heavens of the planets and the "fixed" stars are a number and a harmony. One detail of the proof will suffice as a sample. Because there are seven intervals in the Pythagorean musical scale, and because there were only five genuine planets known when the scale was invented, and because these five with the Sun and Moon make seven, therefore the planets are a musical scale.

Granting the master's fundamental postulates that all

things are contained in the decad, and that everything is number, no logician or mathematician will quarrel with the Pythagorean or Platonic proofs once he has had the patience to transpose them into the symbols with which he is familiar. Some such exercise as this should convince almost anyone that a result reached by the strictest mathematical (or deductive) proof may have no relevance for the world of science and sensory experience, or even for that of common sense. If the postulates are not in accord with verifiable and verified experience, the conclusions deduced from them will have no meaning in the world of the senses. These statements no doubt are truisms, but they are none the less true on that account. Almost any rational man will accept them; yet many rational men ask others to accept factually unverifiable statements because they have been deduced by irreproachable logic, mathematical or other, from assumptions which no rational man need accept. So once again we are induced to soften with forbearance any scorn we might be inclined to feel for the science and theology of our predecessors.

With this in mind it will be particularly instructive to glance at what men whose intelligence was certainly no lower than our own deduced from the recurrent motions of the planets. We shall begin with Plato, as his deductions are the most refined of all.

"When reason," Plato observes, "which works with equal truth both in the circle of the Other and the Same—in the sphere of the self-moved voiceless silence moving—when reason, I say, is in the neighborhood of sense, and the circle of the Other also moving truly imparts the intimations of sense to the whole soul, then arise true opinions and beliefs. But when reason is in the sphere of the rational, and the circle of the Same moving smoothly indicates this, then intelligence and knowledge are of necessity perfected."

All of these statements are susceptible of strict proof, provided we accept the numerological postulates of the Pythagoreans. But none of them has retained any significance, except possibly an extremely rudimentary description of the motions of the planets, it may once have had. Yet not all of the underlying assumptions concealed in Plato's imagery of spheres and moving circles have lost their meaning as the thoughts of men have widened with "the process of the suns." It will be interesting to see what some of the supposed facts responsible for the disguised astronomy in Plato's metaphysics of "sense" and "reason" may have been.

The Pythagoreans inferred from observation—possibly from the contours of shadows during eclipses—that the earth is a sphere or at least roundish. And it was a supposed fact as old as the human race that the stars are affixed to the surface of a vast sphere with the Earth at its center. Here the "sense" —sensory experience—of which Plato speaks led reason astray. No such sphere exists, though the unaided senses report it as certainly as astronomers, extending human vision with man-made aids, report no limit to the depth of "the starry heavens." Firmer than Pythagoras himself in his faith in Number as the All, Plato disdained observation in astronomy and deduced the sphericity of the Earth immediately from the assumption that of all bodies the sphere alone is perfect. Likewise for the celestial sphere of the stars. That both must be spheres followed because the One, the creator of the heavens and the Earth, by the very perfection of his own Being, could fashion no imperfect thing.

The Pythagoreans were less contemptuous than Plato of

induction from sensory experience. At the center of their cosmos they placed Hestia, the Central Fire, to shed light and warmth on the Sun and the planets. This may seem to us a rather poor guess. But we must remember that the gods had to be accommodated somehow, and Hestia offered what was required. Invisible to mortal eyes, the Central Fire was the object of serene contemplation by the immortals, who saw all things but who themselves were not seen. The Father of Gods and Men therefore used Hestia as his watch tower from which to observe erring mankind.

Though Hestia was not the Sun as might be hastily inferred, this hypothetical hearth of the universe gave Copernicus (1473-1543) a hint for his heliocentric theory of the solar system. Or so he (or his officious editor) says in his dedicatory epistle to the reigning Pope of his time, possibly in an attempt to escape censure by endowing his innovation with a respectable antiquity. Thus Hestia, though a pure fiction and a weak hypothesis, at last justified Pythagoras in the court of science.

Round the invisible hearth of their universe the Pythagoreans spaced the Earth, the Moon, the Sun, the five planets known in their day, and the sphere of the fixed stars. The total fell one short of the necessary ten demanded by the perfect decad. We have already seen how they supplied the deficiency by imagining their unobservable Antichthon between Hestia and the Earth. This tenth body of their celestial system was more real to them than all the other nine, for it was number ten.

A skeptic who disbelieved in the gods could hardly accept the theological explanation for the invisibility of the Central Fire. To satisfy him the Pythagoreans invented one of their most ingenious theories. The inhabited regions of the earth, they pointed out, were all on that side of the earth which is always turned away from the center of its orbit. So to view the Central Fire it would be necessary to go beyond India. As not even Pythagoras himself had traveled so far, it was unlikely that anyone else would. But suppose someone did. Hestia would still be invisible to him, because between it and the Earth, Antichthon would cut off the view. Could not the traveler wait till the Counter-Earth rolled by? He could not: Earth and Counter-Earth kept even pace together as they revolved about the Central Fire. Not even the nineteenth-century inventors of the space-filling ether imagined a more satisfying explanation of the impossibility of observing the unobservable.

The celestial decad being now numerically complete, each of the ten celestial bodies was attached to its own revolving sphere. The distances of these eternally gyrating spheres from the Central Fire were then assumed to bear simple numerical relations to one another. Naturally the tetrad and its harmonic ratios were discovered in this celestial arithmetic. They had been deftly insinuated into it before astronomical calculation began. The decad being also implicit in the ten heavenly bodies themselves, the spheres in their motion gave forth an inaudible music—"the music of the spheres"—which charmed scientists and poets from Pythagoras and Plato to Kepler and Shakespeare: "There's not an orb which thou beholdest, But that in its course like an angel sings, Still choiring to the young-eyed cherubim"—as Lorenzo told Jessica. Absurd, no doubt, but somehow less depressing than the nautical almanac.

The crowning ingenuity of it all was the entirely rational

explanation of why it is that mortals—with the occasional exception of a stubbornly imaginative Kepler—hear nothing of the celestial harmony of the spheres. This takes us back to Pythagoras and the legendary anvil. Because it is clanging forth eternally, day and night, year in, year out, the music of the spheres makes no more impression on our exhausted hearing than would the continued din in a smithy with ten musical anvils. That touch must have been added by the master himself.

Perhaps of more rigid intellectual honesty than Pythagoras, Kepler disdained to slip out of a real difficulty by this too convenient subterfuge. Convinced that his ethereal soul if not his gross senses perceived the celestial harmony, Kepler transcribed the song of the spheres on a sheet of music paper. The slower bodies, being nearer the Central Fire, sang bass or contralto as they had for Pythagoras; the remoter and swifter sang tenor or soprano. The melody is scarcely comparable in complexity to any in Holst's curious symphony of the planets. Yet it comforted Kepler as he hummed it to himself while he calculated one misfit orbit after another in the most prodigious arithmetic ever carried through by fallible man to a verifiable scientific conclusion. The music that Kepler heard must have been clearer and simpler than the siren melodies that Plato transposed into his celestial metaphysics.

Ascending now to the more elevated regions of Pythagorean astronomy, we must return for a moment to a remote prehistory, long before Egypt and Babylonia were even imaginable possibilities to the homeless tribes wandering over the future sites of those ancient civilizations. No race of savages in a temperate climate has yet been found so lethargic in intelligence that it is unaware of the invariable repetition of spring, summer, autumn, and winter. Centuries of passive attention to the aspect of the heavens as it changed with the seasons taught the primitive observers that the recurrence of the seasons and the wheeling of the constellations are as predictable as the sequence of day and night. Slowly expanding knowledge of astronomy then revealed more intricate periodicities in the motions of the heavenly bodies and the barely perceptible creep of the seasons, culminating after thousands of years in the awesome discovery of the precession of the equinoxes. This fixed a Great Year in the heavens, at the end of which all the motions would begin retracing the pattern they had just completed, till another Great Year had run its course across the heavens, when the cycle would be repeated once more, and so on for as long as the stars shall last. The repetition in this eternal recurrence is exact to the minutest detail: it is not a mere succession of creations such as Anaximander imagined.

One of the clearest and most suggestive statements of this cyclic theory of the universe, long since abandoned, of course, is that of Eudemus (*c.* 350 B.C.), a pupil of Aristotle and a historian of mathematics and astronomy. His version at least hints at the source of Plato's vision of the Great Year or Eternal Recurrence. Eudemus is speaking to his pupils: "If we may believe the Pythagoreans, I shall once more gossip with you, this little staff in my hand, and again as now will you be sitting before me, and so will it be with all the rest." Time and Eternity are closed, as in the symbol of the serpent Ouroboros, the tail-swallower, the devouring, undevoured "worm that never dies."

This somber dream of the circularity of time is said to have visited the Babylonians when they discovered the precession of the equinoxes. Until then Time had not been wholly swallowed up by Eternity, and there was still hope that man might become the master of his future. But when it appeared that the heavenly bodies repeat all the complications of their motions endlessly, it seemed that Time, about to be overtaken by Eternity, instantaneously united past and present. Thereupon everything that had ever happened began happening all over again precisely as it must have happened time after time since Time and Eternity first became one.

There must be something irresistibly seductive to the imaginative mind in this ancient figure of the circular serpent. A curious feature of its own recurrence in speculative philosophy from the Babylonians to Nietzsche (1844-1900) is that many of those who believe in the circularity of Time also believe they were the first ever to conceive this unending repetition of the universe. That it involves an obvious self-contradiction is nothing in its disfavor for those who, like the unhappy Nietzsche, torture themselves into insanity contemplating the horror of endless reincarnations in their present shape. As an instance of what an undisciplined imagination can do in the way of extrapolation from a single observed fact —here the precession of the equinoxes—the Eternal Recurrence is without a superior in reckless daring.

It would be interesting to unravel some of the numerology in Plato's version of the Eternal Recurrence, especially the threads that snarl up his Great Year with the Nuptial Number in one tight knot, but we must pass on with a brief indication of one or two clues. The Nuptial Number is said by some experts on Platonism to be the fourth power of 60, or

12,960,000. This large number was mentioned in connection with Babylonian arithmetic, and it was remarked then that one of its strongest claims to the thoughtful attention of all numerologists is the plethora of its divisors. The mystical implications of this fact of elementary arithmetic are inexhaustible.

To give a hint of the possibilities we follow Plato and restrict our scrutiny to the two divisors 360 and 36,000. The first is a rather crude primitive approximation to the length of the year. The defect of a little more than 5 days is entirely negligible for the purposes of numerology, and Plato was justified in ignoring it, for the earliest Sumerians or the still earlier savages who long preceded them possibly did likewise. Hence 360 was Plato's Pythagorean or terrestrial year, and 36,000 is exactly 100 such years. But 100 is "justly" the square (10×10) of the divine decad (10), and therefore divinely divine. But again some of the Pythagoreans (and possibly also some of the Babylonian astrologers) asserted (from a first greatly mistaken guess at the period required for one complete cycle of the equinoxes) that 36,000 terrestrial years are exactly equivalent to one equinoctial or cosmic year. From this Plato inferred that the full life of a just man is—or should be—100 terrestrial years, each of 360 days. So it follows, as Plato states, that a day in the life of a man is equivalent to a year in the life of the universe. Man lives only a short span, but he lives that span fast.

Even eschatology, statecraft, and epistemology are included in the Pythagorean synthesis, as of course they must be. Otherwise everything would not be number. A single specimen of each will suffice. The first two are of no historical importance;

the third is. All are typical of the Master's philosophy, and doubtless all were thoroughly expounded in the lost bible of Philolaus.

First there is the somewhat gruesome numerology of Hell as presented by "a brave man, Er, the son of Armenius, a Pamphylian by birth" in Plato's *Republic*. Those interested will find the details there.

Then there is the mysterious number 5040 which Plato gives in his *Laws* as the population of his Ideal City. Anyone who has studied permutations and combinations in elementary algebra will recognize 5040 as the total number of different ways of arranging 7 things in a row, say, 7 books on a shelf. The number is $1 \times 2 \times 3 \times 4 \times 5 \times 6 \times 7$. Written thus its numerological possibilities are embarrassingly evident. Even the super-sacred 7 occurs, to say nothing of the female 2, the male 3, the just 4, the 5 regular bodies, and the perfect 6. Among its other claims to civic attention, 7 is the number of Plato's hills that must be surmounted to attain knowledge and wisdom; in fact 7 "is" actually these hills. But there is infinitely more concealed in this encyclopaedic number. Any cosmic numerologist will observe that 5040 has exactly 60 divisors, while 60 has exactly 12, and 12 has exactly the perfect 6, and 6 has exactly the just 4, while 4 has exactly 3, and 3 has exactly the female 2, which has exactly 2, and so on, 2-2-2- . . . forever. From these facts it can be shown that the Ideal City is contained in the Nuptial Number and that it recurs eternally once it is firmly ("fourly") established. The implications of the Zodiacal 12 are too obvious to need mention. The 3 epitomizes the Ideal Family of the City all through the Great Year.

The third item is of quite a different order of sense. It is the limited principle of dichotomy or division by 2 and, except for the decadic limitation to 10 dichotomies, a fundamental device of classical logic from Aristotle to the Middle Ages and beyond.

We remarked in an earlier chapter that a first step in science is classification, as in the natural history of plants and animals. Dichotomy is one method of pigeonholing a comprehensive class of things in successively smaller compartments or subclasses until, if the process is continued long enough, the original class is split into subclasses containing either no member or exactly one each. At each step at least one of the subclasses is split into two.

Now all things, according to the Pythagoreans, are divided into two categories of contraries, one of which falls on the side of the Limited, the other on the side of the Unlimited. But since the decad is the pattern of the universe, there must be exactly 10 pairs of contraries. Each pair is a dichotomy of the One. For example, no body in the universe (the One) can be both at rest and in motion at any instant of time, and every body is either at rest or in motion at each instant. So Rest and Motion constitute one pair of contraries which suffices to dichotomize the All, the One. The complete decad of contraries was decreed by the Pythagoreans to be the following:

(1) Limited—Unlimited
(2) Odd—Even
(3) One—Many
(4) Right—Left
(5) Male—Female
(6) Rest—Motion
(7) Straight—Crooked
(8) Light—Dark
(9) Good—Evil
(10) Square—Oblong

After even the few examples of Pythagorean science exhibited it should be evident that this dichotomized decad contains unlimited possibilities for the numerology of science, philosophy, and the pure reason. Many of them were elaborated by the numerologists and logicians of antiquity and the Middle Ages. Millions of man-hours and thousands of lives went into this interminable task, to what purpose no man now living can say. All that labor left only a slight residue, in the trivial refinements of a logic that has long been of only antiquarian interest to the few who still are actively conscious of its existence. And while all this apparently sterile field was being cultivated with a fervor almost unique in the history of human thought, the potentially richer field of experimental science, which Pythagoras also marked out, lay neglected and barren.

For this misdirection—if it shall appear in the last reckoning to have been such—of human endeavor Pythagoras is primarily responsible. When he discovered the law of musical intervals two roads lay open to him. One led back into the darkness of myth and superstition; the other led forward into the unexplored possibilities of experimental science. Having set foot on the untried road, Pythagoras abruptly turned back and followed the old familiar road into the past. But he did not go all the way back as did some of his disciples. If in his too ambitious attempt to create a comprehensive science of the universe his thinking was prescientific, the like cannot be charged against his contribution to orthodox mathematics. It was only in his zeal to force everything, from the stars to human values, under the domination of numbers that his thinking became both prescientific and prelogical. In that phase of his intellectual activity he was essentially as far back as the Stone Age. The primitives of that long twilight between total savagery and dawning civilization can have been no less superstitious in the face of nature than was Pythagoras with his wishful dream that the infinite complexity of nature is as simple as a child's arithmetic. And this, as will appear presently, was his own final revelation.

The roads are still open. Which shall we follow? Perhaps it will not have mattered much by the time we reach the end of the journey and lie down to sleep in the chill rays of a dying sun. But whichever way we choose we must be cold indeed if we cannot feel a little of the warmth that cheered the first believers in a rational universe. One thing alone in the eternal flux seemed to remain constant and to be a clue to a unity, to a One, in the chaos of diversity and the confusions of the Many. Two fires, two planets, two parents were merely different aspects of one unchanging number, and the unchanging is eternal. If Number was not actually the Deity, neither had the power to create or to destroy the other. Number ruled the cosmos with an inflexible justice more equitable than that administered by any pantheon of capricious deities. Number alone gave the lie to "Nothing is permanent." It alone existed and endured; all the rest was transitory appearance and illusion. Everything was number in the only sense that gave existence any meaning, and to understand anything, from the motions of the planets to the workings of divine justice, it was necessary and sufficient to understand number.

It was a great dream, as simple and as childlike as it was great, and it has lasted. But it was only a dream, and its recurrence in our own time may be nothing else.

CHAPTER - - - - - - - - - - - - - - 15

Himself Made It?

> "All that we see or seem
> Is but a dream within a dream . . ."

expresses a doubt that many a man of science has felt in reviewing the outcome of a lifetime's struggle to understand his relation to the universe. So it was with Pythagoras, if we can credit the testimony of the disciple—whoever he may have been—who recorded his master's last earthly dream. It was a sufficiently terrifying dream, yet not without a certain grandeur characteristic of the dreamer, and one that has returned in a less sinister shape to trouble the foremost Pythagoreans of the twentieth century. One modern version may be given first, as a contemporary introduction to its ancestor of twenty-five centuries ago. Pythagoras himself might be speaking; but it is Eddington, in his *Space Time Gravitation* of 1920.

"It is one thing for the human mind to extract from the phenomena of nature the laws which it has itself put into them; it may be a far harder thing to extract laws over which it has no control. It is even possible that laws which have not their origin in the mind may be irrational, and we can never succeed in formulating them. . . .

". . . we have found that where science has progressed the farthest, the mind has but regained from nature that which the mind has put into nature.

"We have found a strange foot-print on the shores of the unknown. We have devised profound theories, one after another, to account for its origin. At last, we have succeeded in reconstructing the creature that made the foot-print. And Lo! it is our own."

When he felt the need of solitude, Pythagoras would retire with his lyre to that same Grotto of Proserpine where Theano had come to him. Meditating and strumming his lyre there one day, he struck out the chord of the major third, and passed with the dying music into his dream.

All that "the stern grip of necessity"—his own phrase—had prevented him from doing in his waking life now became easy to him. Awake, he had resolved music into number; in his dream, he found empirical keys as simple as his monochord for unlocking all the secret doors of the physical universe. From the motions of the heavenly bodies to the gyrations of the ultimate particles of matter, some intelligence more penetrating than his own guided him to the few and astonishingly simple laws governing all. Like his own law of musical intervals, the laws of the entire universe apprehended by the senses were mathematical. Therefore they could only be the work of mind. But of whose mind? Only the mind of the Great Architect of the Universe, the Supreme Mathematician, could have thought these few simple and universal laws governing everything from planets to atoms.

Pressing on to the limit of the universe, the dreamer came

to the impenetrable wall of Chaos—"the Outer Unlimited"—bounding the knowable.

The dream now became less vivid. Slipping back into one of his earliest incarnations, Pythagoras dreamed that once again he was Orpheus. In that shadowy life he had charmed all living things with his music, and even the inert rocks had responded to his lyre. Now, in the dream within his dream, he saw that everything in the universe obeyed his music. He, not the Supreme Mathematician, was master. With this prophetically disturbing recognition the dreamer ceased to be Orpheus and, descending to a deeper dream, became again the human seeker after ultimate Knowledge.

The Orphic episode was but a minatory interlude, forewarning the presumptuous dreamer of what was to come should he continue dreaming. The chord of the dominant seventh recalled him to a dream-world of the human senses. There Pythagoras, his powers of perception superhumanly heightened, resumed his analysis of material things. This time, doubting its existence, he would seek and surely find the ultimate reality.

The geometrical forms of all bodies became familiar to him; and as he passed down the scale from the tangible to the intangible, from things of the senses to the invisible atoms composing them, he found the same mathematical laws governing all. Continuing beyond the atoms, he dissipated all matter. Only thought, manifested as mathematics, persisted. He began to doubt the existence of an outer bounding wall between reason and Chaos, between the existent and the non-existent, between Being and not-Being.

If there was no longer a boundary to the knowable universe, where was he? Could the wall, too, have vanished into nothing or into pure thought? If anything at all had resisted his reason, the infinitely hard, impenetrable wall of Chaos might remain unchanged for him to touch. This impalpable nothing that streamed through his fingers, this futile residue of all his analyses of things of the senses, filled him with fear and an uncontrollable desire to strike his hand against something unyielding. Impelled by a horror of madness he began groping through the void in search of the bounding wall of Chaos.

It moved toward him. But it was not as it had been when first he imagined he saw it. Now it was an infinite mirror without substance. He seemed to meet himself in the mirror and to become one with an unreal other. United with his own image, he passed into the wall. What had seemed impenetrable Chaos offered no resistance.

Beyond the Chaos of the Outer Infinite bounding the universe, where was he? Surely he had passed this way before? He realized that the universe was closed, but not by Chaos or any wall, for he was returning to the point from which he had started. But where was he now?

With a shock that jarred his reason he recognized that he was at once everywhere and nowhere in a contradictory universe of his own thoughts. Struggling to find something outside himself to which his mind might cling, he remembered his disciples. Then he doubted their existence with the rest. Were they, too, the insubstantial creation of his own mind?

To reassure himself he repeated their formula of finality, "Himself said it." All their doubts about the nature of reality were silenced by that formula. As he woke at the sound of his own voice, Pythagoras heard himself admonishing them, "Say rather 'Himself made it?' "

So saying, he rose and walked out of the Grotto of Proserpine to face his living disciples. He met them with "a grim and ghastly countenance, and said that he had been in Hell."

The peaceful days in Croton were almost at an end. Two deep-seated ills in the Pythagorean philosophy had ripened and were about to precipitate the dissolution of the Brotherhood as an organized body. One has already been noted: the aggressive antidemocratic philosophy of the Brotherhood and its persistent meddling in politics incurred the sullen hatred of the greater part of the population. The dispute over the division of the spoils after the conquest of Sybaris gave Cylon his opportunity to stir this passive hostility to active revolt against the aristocratic oligarchy. The graver weakness of the organization was its secrecy.

The mysteries most jealously guarded by the Brothers were of supposed cosmic and practical importance. It is both tragic and ironic that they were preposterous nonsense. A major cause of the revolt against the Pythagoreans was the fear that this esoteric knowledge wielded by the aristocrats would remain forever in their hands. Rumors of potent mysteries being hoarded by the mathematicians leaked out to the democratic mob. Malcontents of all factions and of all shades of unintelligence put their own constructions on the distorted hints of what Pythagoras had intoned behind the curtain. Why should only the mathematicians be permitted to share their master's most dangerous knowledge? Because they were his familiars to execute his nefarious purpose in profane necromancies. The mob began howling that it was being oppressed by magic. Pythagoras and his chosen few were undermining the foundations of democracy with their occult deviltries. Away with them before it was too late.

That part of all their teachings which the Pythagoreans themselves esteemed as of the highest worth was also so regarded by the mob. The one thing of all the master's inventions that might have helped the downtrodden to rise and the slave at last to shed his chains was passed over as a frippery of no human value. If the rebels against good government as Pythagoras understood it ever suspected the existence of experimental science they made no effort to acquire it or to control its use—as today the rebels against bad government overlook their most obvious opportunity. Instead they hankered after such powerful aids to mental and physical well-being as the knowledge that five is the male marriage number and six the female; that two is opinion and four justice; that the deity is the number one; that a quart pot should never be sat on; that a fire should never be stirred with an iron poker; that a mirror should never be looked into while it is reflecting a light; and that, most important of all, the beam of a balance should never be stepped over, lest the scales of divine justice be tipped hellward. Offered science, they chose superstition.

As the master had expounded the sacred mysteries they had not seemed utterly childish and devoid of meaning. But anyone who so willed might take the most rational of them and with but little imagination parody it into an absurd travesty of sane reasoning. By theft Cylon had obtained an exposition of all the numerology of the divine decad. To allay the apprehensions of his followers he shared with them all the higher mysteries of the mathematicians. But as he caricatured the secret doctrines to his mob in a frenzy of ridicule and scathing contempt they sounded like the meaningless meanderings of a lunatic.

The mob lost its consciousness of ignorant inferiority. Cheered, it began to share the demagogue's simulated superiority to all knowledge and wisdom. Individually they still feared the magic of the mathematicians. But as a mob they were easily harangued into a cowardly sort of courage and sensed that they, the many, were in spirit one. They were now ripe for collective murder. Wherever Cylon might lead they would follow. He led them to Milo's house.

Stronger than any twenty of the maddened mob, Milo fought his way out and escaped. Years later, when he was an old man in bitter exile, he met his end alone. While attempting to split a forked tree with his hands his strength deserted him. The crack snapped shut, and his hands were caught as in a vise. It had been snowing. By day the struggling figure in the snow was starkly visible, and when the full moon rose the naked landscape shone out as bright as day. Before the helpless man could starve and freeze to death the wolves found him.

Others did not escape. Accounts differ on the number of those who died in the flames. Some say Pythagoras himself was trapped with his foremost disciples when the mob fired the house after blocking every exit. Others say he escaped and made his way to Metapontum, where the people honored him and his teachings till eternity claimed him. Still others say that he did not wait for time to overtake him, but on reaching his seventieth birthday, decided that he had lived long enough and abstained from food and drink. Two centuries, some say less, after his death the Pythagorean Brotherhood had ceased to exist anywhere.

CHAPTER - - - - - - - - - - - - - - 16

Intimations of the Infinite

TWO consequences of lasting significance for science and philosophy developed from the positive achievements of the Pythagoreans in arithmetic and elementary geometry.

The first was the belief that "number" may be so defined that at least the physical universe can be consistently described in terms of "numbers." The second was the common belief that conclusions reached by mathematical reasoning have a greater certainty than those obtained by any other means. Both have been questioned, especially since the last decade of the nineteenth century. Each was successively modified many times to accommodate increasing knowledge, but the basic assumption in both remained substantially unchanged. Together they still are complementary postulates of one as yet unverified hypothesis: a rational account of (at least) the physical universe is possible which, when finally given, will agree with sensory experience and empower human beings to predict the course of nature.

It is not generally anticipated in this ambitious dream that the whole of nature will ever be summed up in a single formula, as the numerologists of antiquity believed it was. But the vision of increasingly more inclusive syntheses and successively closer approximations to "reality" is not universally regarded as an illusion, though doubters are not uncommon. From past experience it appears that the first step into the unknown in a particular direction may take us far, the next not so far, and so on until our race is extinct. But the sum of all will approach an imaginable finality, though never reaching it, and few will have to be retraced.

Neither in mathematics nor in science has any certainty of such steady progress been attained. The hope is that by continuing in the way we have come so far, we or our successors may find the sure road to the future. The record up to the present is a confusion of tentative explorations in many directions with frequent returns almost to the starting point, but not quite. Some notable advances have been made, if only in the discovery of obstacles unsuspected by our predecessors. Just as they either removed the obstructions in their path or by-passed them and proceeded in new directions, so may we.

In this continual though slight progress, each epoch passes on to the next a moral obligation not to neglect any problems that have been left only partly solved. So long as the past is not clear the future is uncertain. Two obstinate difficulties of twenty-five centuries ago still resist complete resolution and continue as prolific of new methods in rigorous thinking as they were when first encountered. One concerns the meaning of "number"; the other, the scope and reliability of deductive reasoning. Both originated in the optimistic belief of the Pythagoreans that "number" as they first knew it was the simplest language adequate for both mathematics and a rational description of the entire universe. It was too simple and infinitely inadequate.

As we have seen, the early Pythagoreans recognized that

the natural numbers 1, 2, 3, . . . and the fractions or "ratios" generated by dividing one whole number by another are not the species of "number" required to measure so elementary a "magnitude" as the diagonal of a square whose side is one unit in length. More concisely, they proved in effect that the square root of 2 is not a rational number. They had supposed in particular that their original "numbers"—the rational numbers—would suffice to measure any line. Then arose the question of the meaning of the "length of a line." Were all lines necessarily measurable by numbers? Three possibilities lay open. Either irrational numbers—such as the square root of 2—might be denied the status of "number"; or the original restricted concept of number might be widened to include both rationals and irrationals; or a totally new start might be made in which numbers are not explicitly correlated with lines. The Greeks after the Pythagoreans chose the last, and in so doing were confronted with the mathematical infinite.

To reason at all about the infinite they were compelled to refine their deductive technique. Temporarily overcoming the particular difficulties presented by irrational numbers they inadvertently admitted subtle assumptions into their logic. These either passed unnoticed or were ignored as irrelevant for mathematics until they made themselves acutely felt in modern mathematics toward the close of the nineteenth century. Inconsistencies and paradoxes then began appearing in the very foundations on which a major part of all mathematics since the seventeenth century had been reared. All were traced first to an imperfect understanding of the mathematical infinite. Closer analysis of some of the paradoxes of the infinite next showed that the more serious difficulties had been concealed for centuries in the logic which the great mathematicians from ancient Greece to the late nineteenth century had supposed adequate for the development of mathematics.

Some of these logical flaws were strangely unmathematical at a first glance. One was of the same type as the remark of Epimenides the Cretan that "all Cretans are liars." Others will be noted as the occasion may demand. For the present it will be sufficient to see how the ground for these unwelcome weeds was cultivated by the mathematicians and logicians in the interval between Pythagoras and Plato.

It is plain from his writings that Plato was vividly conscious of the fundamental difficulties for epistemology presented by irrational numbers. His struggle to overcome them may have been partly responsible for his alleged abandonment of his theory of Ideal Numbers. In his old age, Plato is said by some to have believed that his capital achievement, his theory of Ideas, if not wholly untenable, was unworkable. Whether or not this is the fact, it is significant that one of the greatest philosophers in history should have thought it worth his serious effort to understand the nature of number, particularly of irrational numbers. The problem of irrationals moved him strongly, and he berated his fellow Greeks for still believing—in the majority—with the Pythagoreans that all "measures" are rational. "Whoever does not know that a diagonal of a square is incommensurable with one of its sides," he declared, "is not a man but a beast."

The primitive mind seems to have an instinctive horror of the infinite—the unterminated, the unbounded, the unlimited—in any of the numerous shapes in which it forces itself on the attention even of savages. The familiar objects of their daily lives would seem static and essentially unchanging, each with its own permanently recognizable individuality. The tree here in this place today would not have vanished tomorrow. Though it might be alive, and no doubt harbored a spirit, it was the same tree and not another from day to day. But the wind was dynamic, never the same from instant to instant, changing in strength and direction continuously. It was beyond all human control—"the wind bloweth where it listeth" —and its comings and goings were inaccessible to human foresight. In some sense it was more alive than the rocks and trees; for it was in no way limited either in space or in time. Ages later, when men had learned to count freely and fearlessly, things which were limited in extension like pebbles and trees were brought under the rule of number and were counted. But the winds and the continuously flowing streams escaped this domination. Whatever it might be that enabled them to pass from one place to another, and to remain the same all through the interval, could not be counted. Motion eluded number. It was unlimited, unbounded, infinite, not one, nor yet a many in the same sense that a handful of pebbles is a many.

But long before this intimation of the uncountable infinite was perceived instinctively, a lesser though still sufficiently disturbing infinite had emerged from the countables of nature. The "natural numbers" that enumerated the rocks and trees appeared to have no end, though the tangible things counted by the numbers might all be comprised in a single collection. What were the numbers counting when all the things in the world, and all the stars in the sky, had been enumerated? The mind could not conceive a limit to the numbers, a greatest of them all which no number could surpass, though it might easily imagine an end to the totality of countable material things. What then remained in the universe for the self-generating numbers to count but the numbers themselves? Nothing; the numbers must have an existence independent of things. So the Pythagoreans imagined, and after them all who have believed that numbers were not invented by human beings but were discovered and are merely observed.

Some of the foremost mathematicians of the present and near past, refusing to commit themselves to this sharp dichotomy, have compromised on intermediate positions. For Gauss (1777-1855), usually included among the three or four greatest mathematicians in history, number alone of all mathematical concepts was a necessity of rational thought, if not actually "a creation of the mind." For L. E. J. Brouwer (1882-   ), a leader in the revision of the logic of the infinite, human beings are born with "an original intuition" of "an unending sequence of individually distinguishable objects," and therefore, it would seem, are endowed at birth with the capacity for imagining that the sequence of natural numbers has no end. But for the majority there is no middle ground. Numbers either are human inventions or they exist "out of space, out of time," as Plato's Ideal Numbers existed for him, forever independent of the human mind though not beyond the feeble grasp of human thought.

Whoever first perceived that the natural numbers have no end must have been crushed by the sudden revelation. His finitely countable days, even if he should live a million years, were as nothing in the endless ages of eternity, and all his life was but a momentary flash in an infinite darkness.

Something of that forgotten horror survived in the decad of the Pythagoreans. To evade the "countable infinity" of Number confronting them, they took refuge in the fiction that all numbers beyond the primitive ten of finger-counting have only a repetitive, imitative reality and can be ignored for the purposes of science and philosophy.

The earliest documentary evidence that this superstitious horror of the "countable infinite" had been overcome is Euclid's proof (about 300 B.C.) that the sequence of natural primes 2, 3, 5, 7, 11, 13, 17, 19, 23, 29, 31, 37 . . . is endless. The proof is indirect and could have been imagined by the Pythagoreans themselves, had they not been too timid to trust their reason beyond the finite evidence of their sensory experience.

Concealed in Euclid's exceedingly ingenious proof is a hint of the subtle logical difficulties which came fully to light only in the twentieth century. The most important of these concern the indirect method of proof and the meaning of "existence" in mathematics. To describe their nature it is necessary to recall two details of traditional deductive reasoning. These also will appear later in connection with Plato's dialectic.

If we hope to prove that a certain statement S is true, and if no other way suggests itself, we assume, on the contrary, that S is false. Then, if from this assumption we can deduce a contradiction, it follows immediately in classical logic that S is true. This is the "indirect method of proof," the familiar *reductio ad absurdum*, or reduction to an absurdity, of school geometry. Euclid's first use of it is to prove that if two angles of a triangle are equal the sides opposite those angles are equal. He also resorted to it in proving that the sequence of primes is endless.

Another device of classical logic also is of frequent use in mathematical reasoning. Instead of assuming as in the indirect method that the *S* which we hope to prove is false, we assume that it is true. We then deduce consequences of this assumption. If one of these is known to be true, and if the steps which led to it are logically reversible, then we can infer by the rules of classical logic that *S* is true. But if the steps are not reversible, we cannot infer the truth of *S*, and indeed *S* may be false. The necessary reversibility of the steps is sometimes overlooked by hasty or incautious reasoners. This method has been called "analysis"—though the word has another important meaning (unnecessary for our purpose) in modern mathematics. It is attributed by some historians to Plato, who certainly appreciated its power in both philosophical and mathematical reasoning, even if he may not have been the first to invent it or to advocate its use in geometry.

The indirect and the analytic methods together constitute the major tactic in at least the earlier stages of Plato's "dialectic"—a hard word to define concisely, but meaning, not too vaguely, a method of reasoning capable of discovering truths. In dialectic, falsities are pared away by argumentation until either nothing or only a nucleus of rigidly demonstrable assertions remains. Once more, however, the nature of the truths discovered depends upon that of the postulates on which the logic operates. The postulates themselves may be granted universal validity by the logician, and the logic may likewise be conceded infallibility. The outcome then is a system of truths acceptable to those who agree that both the postulates and the logic are unobjectionable. In particular, if the system is to satisfy a rational mind, the logic must be incapable of producing inconsistencies from the postulates to which it is applied. It is at this point that modern mathematicians have found it necessary to proceed with caution.

A statement about a finite collection of things can be proved or disproved either by testing it for each of the things in turn or, if the things are too numerous, by giving an explicit rule whereby such testing could be carried through in a finite time. If the things are statements and it is required to establish the truth of all of them, classical logic permits us to assert that each of them is a definite one of "true," "false," and the test must be capable of deciding which. Again, each of the things in the finite collection has a recognizable individuality by which it can be distinguished from each of the others; the thing is itself, not something else. We are still within the domain of common sense, and so far nobody has seriously objected to mathematical reasoning about finite collections based on these assumptions of traditional logic. But with infinite collections there is occasion for rational doubt.

Suppose for example that an arithmetician asserts that every natural number is either even or not-even. As the collection of all natural numbers is endless, it is impossible to test each of them (by dividing it by 2 and noting whether the remainder is 0 or 1) to establish which it is. Likewise for the primes: any natural number, we assert, is either prime or not-prime and, given any specific number, we can decide which in a finite number of humanly performable operations. But if we cannot produce all the even numbers or all the primes, in what sense if any is it meaningful to state that all natural numbers are either even or not-even, prime or not-prime? And in what sense does anything which cannot be produced or exhibited by a finite number of performable operations exist? Does a proof which demonstrates the "existence" of a certain thing, without specifying any method for constructing it, have the same logical reliability as one which actually shows how to produce the "existent" thing?

Such doubts do not perturb those who believe that numbers have an independent existence of their own, and that human beings only observe the ideal realm in which numbers will continue to exist when the human race has ceased to cumber the earth. Likewise for the rules of classical logic and the theorems of geometry; these too "exist" in the extra-human sphere of Eternal Being.

Others, of an earthier temperament, seeking to discover any inherent limitations to which a specific system of deductive reasoning may be subject, reach such unexpected conclusions as the following. In any deductive system inclusive enough to take in the arithmetic of the natural numbers, "undecidable" statements may be constructed. A statement is said to be "undecidable" in a particular system if neither its truth nor its falsity can be proved by means within that system. The existence of undecidable statements is demonstrated by exhibiting them and proving that they are undecidable. It is not a matter of being unable to prove or disprove certain statements for mere lack of ingenuity. An undecidable statement will never be proved or disproved by anybody.

This is the finite kind of certainty that has emerged from about twenty-three centuries of deductive reasoning from Plato and Aristotle to Gödel, who first constructed (1931) an undecidable statement. The philosophers of antiquity and their orthodox followers in the Middle Ages appear to have striven after an omnipotent logic that would ultimately settle every question either affirmatively or negatively. The mathematical logicians of the twentieth century have shown that, in mathematics at least, the ancient goal is unattainable. But the efforts of all the mathematicians and logicians from Thales to the twentieth century to attain the unattainable were by no means a mere waste of time and thought. Originating in the recognition by Thales that deductive reasoning is both possible and profitable, and continuing in the successful efforts of the Greek mathematicians between Pythagoras and Plato to give a consistent account of both rational and irrational "magnitudes," the quest for universal certainty uncovered much of lasting interest for philosophy no less than for mathematics. Centuries later some of what was thus discovered in the cultivation of knowledge for its own sake proved indispensable to the lonely workers in the dawn of modern science. To cite a classic instance, Kepler probably would never have recognized the planets' orbits as ellipses (with the Sun at one focus) if the Greek geometry of conic sections had not been available for his use. Without Kepler's laws of planetary orbits as a guide, Newton might never have proposed his law of universal gravitation; and without that, the development of astronomy, the physical sciences, and modern technology would have been quite different from what it has been in the past two and a half centuries.

The devastating discovery by the Pythagoreans that not all numbers are rational (that is, of the form a/b, where a, b are integers) marked a major turning point in the development of deductive reasoning. It was one definite beginning of the mathematical theories of continuity and the infinite.

It was also the occasion for much new epistemology and the revision of some old; and in the direction of modern science the resulting Greek theory of continuity prepared the way for an understanding of motion. This epochal landmark in the progress of mathematical and speculative thought is so outstanding that something of its history may be of interest in passing.

After the discovery that the square root of 2 is not a rational number, the Greek geometers proved the like for many other square roots. By Plato's time the existence of irrational numbers (as the matter would be phrased today) was engaging the attention of philosophers who were only incidentally concerned with mathematics. In Plato's dialogue *Theaetetus* Socrates tries to make Theaetetus say what knowledge "is": "Take courage then and nobly say what you think knowledge is." Taking his courage in his hands, Theaetetus replies, "I think that the sciences which I learn from Theodorus [of Cyrene, flourished 380 B.C.]—geometry and those you just now mentioned—are knowledge. I would include the art of the cobbler and other craftsmen. All of these are knowledge."

It is plain that the generous Theaetetus has included too much in his catalogue to please a relentless cross-examiner like Socrates, and the philosopher forces his victim to admit that he has not stated what "knowledge in the abstract" is. Socrates then tries to drag out of him what clay "is." He seems to be wrestling to make Theaetetus grasp the universal Clay—not this clay or that clay—as an Eternal Idea, a Form in which the mere particular clays of brick-makers, oven-makers, potters, and other craftsmen in some sense "participate." Socrates is interested in none of these. He is seeking the universal, the abstraction, the Idea, and Theaetetus rather optimistically thinks he sees the point. In answer to a polite request by Socrates he shares his enlightenment: "Theodorus was writing out for us something about [square] roots, such as the roots of 3 or 5 feet, showing that in linear measurement (that is, according to the sides of the squares), they are incommensurable with the unit. [In our terminology, the square roots of 3 and 5 are irrational numbers.] He selected the numbers which are roots up to 17, but he went no farther. As there are innumerable roots, the notion occurred to us of attempting to include them all under one name or class." Theaetetus tells Socrates that they found the desired classification, but admits that he is unable to give Socrates an equally satisfactory answer about Knowledge, thus justifying Plato's contention—reiterated in various forms throughout his writings—that philosophy is more fundamental and harder than mathematics.

Incidentally, there is nothing in these disclosures of Theaetetus to substantiate the inference of some historians of mathematics that Theodorus of Cyrene was the first to prove that the square root of 2 is irrational. Euclid's semi-geometrical demonstration (*c.* 300 B.C.) is given in Book 10, Proposition 27 of his *Elements*. Though less perspicuous than the strictly arithmetical proof current today, it is more suggestive historically. It exemplifies the radical transformation of Greek mathematical thought consequent on the appearance of irrational numbers. As Euclid states the theorem, "A side of a square and its diagonal have no common measure." Here "measure" is the important word. If a diagonal of a square whose side is one unit in length is not measurable by a "number"—rational number that is—what "is" it? The Greek geometers called it a "magnitude," and constructed a theory of the "measurement" of magnitudes in which, instead of appealing to the familiar natural numbers for sanction, they invoked spacial intuition. Rather than the Pythagorean declaration that "space is number," the new creed might have asserted that "number is space."

It was mentioned in an earlier chapter that geometry must start from certain unanalyzed but accepted primitive concepts, such as "point" and "line." Though a Greek geometer might have attempted to explain what he meant by a "magnitude," it was a primitive notion when he began actually doing geometry. He postulated, though not explicitly, that magnitudes of "the same kind"—say lengths of lines, or areas of plane figures, or volumes of solids bounded by plane faces—can be compared with respect to the relations of equality and inequality. Thus it was meaningful to speak of one magnitude being greater than, equal to, or less than another magnitude of the same kind. A magnitude contained a whole number of times in another was called a "measure" of that other. For example, if the magnitudes concerned are segments of straight lines, or briefly, lines, the line A is a measure of the line B if A can be stepped off some exact number of times on B. If A is a measure of both B and C, A is a "common measure" of B and C. If two magnitudes have one common measure they have any desired finite number of common measures, all of which are constructible from the first. To illustrate, lines 10 and 12 feet long have the common measure 2 feet, and any proper fraction of 2 feet is also a common measure. But a side and a diagonal of a square have no common measure. The Greek geometers expressed this by saying that the diagonal is "incommensurable" with the side. Any magnitudes are called "incommensurable" with each other if they have no common measure. A famous pair is the diameter and the circumference of a circle.

The Greek solution of the problem of measurement hinged on the cardinal definition of "proportion," said to have been devised by Eudoxus. In Euclid's *Elements* this famous definition is the fifth in the fifth Book. We shall state it in its classic form, to illustrate an earlier remark that subtle assumptions slipped unnoticed into mathematics in spite of the utmost care to keep them out. First we note that a "multiple" of a magnitude is a legitimate enough concept: if the "multiplier" is the natural number m, the m-multiple of the magnitude A is constructible by stepping off A m times on a line of sufficient length. If the line is not long enough to accommodate this multiple, it may be produced—lengthened—until it is. The Greek geometers noticed the necessity of including this possibility of producing a line to any finite length as a postulate and did so. It is rather surprising then that they overlooked the infinitely greater necessity in the following definition of "same ratio." "The first of four magnitudes is said to have the 'same ratio' to the second that the third has to the fourth, when, any equimultiples whatever of the first and third being taken, and any equimultiples whatever of the second and fourth being taken, the multiple of the third is greater than, equal to, or less than the multiple of the fourth, according as the multiple of the first is greater than, equal to, or less than the multiple of the second." This defines "same ratio," from which "proportion" follows by a mere verbal definition. "If the first of four magnitudes has the same ratio to the second that the third has to the fourth, the four magnitudes are said to be proportionals, or in proportion."

Such was the verbiage through which generation after generation of schoolboys attempted to grasp elementary geometry before Euclid's *Elements* was discarded as a textbook. It is not necessary for our purpose to translate the definition into readily intelligible form in the symbolism customary today. Even without understanding the simple meaning of the definition, anyone can see as a mere exercise in reading that the words harbor a tremendous assumption in the twice-repeated phrase "any equimultiples whatever." "Equimultiples" of two magnitudes means "the same multiples," for example, three times, or eight times each of the magnitudes. To ascertain whether four magnitudes are in proportion, the "any whatever" of the definition demands that all pairs of equimultiples be tested. As there is an infinity of these pairs the test is humanly impossible. But is this a valid objection? Not for those who can imagine themselves performing an infinity of multiplications and comparing the results as required by the definition. Whichever side may be the more rational is a matter of opinion, unless it should turn out that one or other is led into inconsistencies by its preference. But it is revealing to find that in attempting to avoid the snares of "number" by appealing to the geometrically (or visually) intuitive concept of "magnitude," we lose ourselves in the same infinite as before.

The theory of measurement and comparison of magnitudes was capable—with certain amplifications—of giving a rational account of continuous motion. But, as has often been observed, the Greek genius was unsympathetic to the fluent and dynamic, preferring to impress itself on sharply distinct objects, each marked off from all the others by its finite completeness and perfection. In their geometry this predilection for the static as opposed to the dynamic produced a multitude of special theorems with no hint of a general principle unifying any considerable number of them in one synthesis. Modern geometry is only passively interested in individual theorems. What it seeks and finds are comprehensive generalizations from which any desired number of special theorems can be obtained by uniform processes. The distinction between the ancient approach and the modern has been compared to the difference between chiseling away a granite boulder a chip at a time and blasting it to fragments with a charge of dynamite. Another common simile likens Greek mathematics to the Parthenon and modern mathematics to a Gothic cathedral. The temple is an end of everything it represents, the cathedral suggests no cramped finality.

Whatever may be the justice of these contrasts, or whether they are founded on more than fancy, the Greek mathematicians stopped short of the rational description of motion for which their theory of measurement might have sufficed. Having surmounted the central difficulty by creating a workable theory of both commensurable and incommensurable magnitudes, they halted before a paradox they might have by-passed. Perhaps not fully realizing what their theory implied, they had actually created (or discovered) the continuum of "real numbers" represented spacially by the uncountably infinite set of all the points on a line. But because all their troubles with irrationals had stemmed from the Pythagorean attempt to attach rational numbers to lines, the creators of the continuum deliberately abstained from assigning "numbers" to magnitudes. Lines might be compared with respect to equality and inequality, but a general arithmetical definition of "length" applicable to all lines was meticulously avoided. Until the somewhat nebulous concept of "magnitudes" should be replaced by a generalized and precise equivalent in terms of numbers, a usable theory of motion was scarcely feasible.

Before glancing at the kind of paradox which stopped the Greeks on the very threshold of modern mathematics we may note how Plato attempted to unify all numbers. The Pythagoreans had generated all natural numbers from the One or the Monad by the mystical union of the Odd and the Even, or what was numerologically equivalent, by the marriage of the Limited with the Unlimited. With the discovery of irrationals the Pythagorean categories of Odd and Even, Limited, and Unlimited were no longer adequate to specify either "number" or "space." Instead of Number being in essence discrete, like a handful of pebbles, it was now essentially a continuum, like the atmosphere as reported by the senses. In this inseparable and uncountable whole the natural numbers and all other rational numbers were more sparsely scattered than the stars against the black of midnight. Desiring a unified substitute for the beautiful simplicity of the Pythagorean "Everything is number," Plato sought an extended definition of Number which would comprehend both rationals and irrationals and which, moreover, would include them as numbers with no reliance on spacial intuition as in the "magnitudes" of the mathematicians. Had he succeeded he might have anticipated at least a part of the modern theory of the continuum.

Instead of the Pythagorean Limited and Unlimited, Plato invented and used the "Great-and-Small," which seems to have resembled our continuum—as, for example, all the numbers "corresponding to" the points on a line. From this and the One he attempted to derive his Ideal Numbers which some interpreters, including Aristotle, claim are identical with his Ideas or Forms. But like all his contemporaries Plato was handicapped by the lack of a symbolism capable of capturing and stabilizing the elusive concept he may have imagined, and agreement as to what this might have been has still to be reached by professional Platonists. Possibly if the Ideal Numbers had been a creation of Plato's youth instead of his old age they would have been easier to understand.

We shall consider next the part Zeno's paradoxes may have had, and probably had, in the failure of the Greeks to proceed from their theory of magnitudes to a generalized arithmetic capable of describing motion. The "infinite numbers" implicit in the "magnitudes" of the geometers eluded mathematicians and philosophers alike till the last third of the nineteenth century.

CHAPTER - - - - - - - - - - - - - 17

A Miscarriage of Reason

YOU cannot get to the end of a race-course, because before you traverse the whole course you must traverse half of it, and before you can traverse that half you must traverse half of it, and so on indefinitely. It follows that in any given space there are an infinite number of points. You cannot touch an infinite number of points one by one in a finite time."

But athletes do get to the ends of race-courses, and some of them run a hundred yards in about nine and one-half seconds, which certainly is a finite time. Not only do runners reach the ends of their courses, but the fastest overtakes any who may be ahead of him near the finish and wins the race. There must be something wrong with our eyes, for "the slower will never be overtaken in his course by the faster, since the pursuer must always come first to the point from which the pursued has just departed. The slower will therefore be always ahead."

Still more remarkable, it is impossible to commit murder by the use of arrows, firearms, knives, or any other material implement. For the arrow or the bullet or the knife must penetrate the victim's body, and to do so must move. But it cannot move, because motion is impossible, as may be demonstrated by the same kind of reasoning as before. Yet thousands of men have been shot or stabbed to death, and others have been hanged for the corresponding murders. Either there has been a serious miscarriage of logic or a more serious miscarriage of justice. But there cannot have been a miscarriage of logic, because it is the surest of all aids to the pure reason. Therefore our senses, as usual, must have deceived us. All those races we imagined we saw fleet runners winning, and all those killings we read about in the newspapers were just so many illusions of our sensory experience. They never happened.

If the last sounds like the travesty of sane reasoning and saner experience which it is, we may remember that it can be matched, not once but many times, by equally absurd travesties of sanity and common sense in the historical record. To cite an instance from which all occasion for controversy evaporated long since, we recall that the orthodox logicians of Galileo's day (near the turn of the sixteenth century) rejected the evidence of their senses in the matter of falling bodies. They saw the one-pound shot and the ten-pound shot dropped from the same height at the same time strike the ground simultaneously. But their intuitive logic had required them to believe that the heavier shot must fall ten times faster than the lighter. What they observed must therefore be a deception foisted on the reason by the senses. They proceeded to prove that this was so, and to them it therefore was so. Some of the less cautious then accepted Galileo's invitation to view the satellites of Jupiter through his telescope. They easily disposed of what they saw as purely imaginary bodies generated by imperfections of the glass in the lenses. The Greek astronomical system had provided no accommodations for these factitious satellites. Therefore they could have no real existence. The spots on the Sun which Galileo next induced the rasher logicians to inspect through his "glazed optick tube" likewise were reasoned out of existence. The Sun, it had been known since the time of Pythagoras, was a perfect body. Therefore it could show no blemish. The only consistent logicians among them all refused to look through the telescope. If the reason is infallible, why appeal to the senses at all? Their faith was as great as their lack of common sense.

The paradoxes of the race-course are two of Zeno's on motion. Little is known of the life of Zeno of Elea (flourished, 475 B.C.) son of Teleutagoras, and not much more about his purpose in inventing his immortal paradoxes. Those on motion are perhaps the most popular of the eight which Zeno bequeathed to generation after generation of logicians and mathematicians. Two others on motion are "the arrow" and "the stadium." "The flying arrow is at rest. For if everything is at rest when it occupies a space equal to itself, and what is in flight at any given instant always occupies a space equal to itself, the arrow cannot move." "The stadium" is more difficult to understand without explanation, so we pass it. It lands us in the temporal absurdity that half a given time is equal to the whole time. The other four paradoxes are equally exasperating, but the three stated are sufficient for our purpose.

As there has been much speculation on Zeno's object in devising his paradoxes, we may quote what Zeno is alleged in Plato's *Parmenides* to have said himself on the matter. The doubtful legend of the *Parmenides* pictures Socrates, as a young man of about twenty, meeting Zeno, "then nearly forty years of age, of a noble figure and fair aspect. In the days of his youth he was reputed to have been beloved by Parmenides." Socrates and his friends had "wanted to hear some writings of Zeno, which had been brought to Athens . . . for the first time. . . . Socrates was then very young, and . . . Zeno read them to him in the absence of Parmenides. . . ." So much for the meeting of Socrates and Zeno, which may only have been an invention of Plato's to give his excessively abstract *Parmenides* a touch of humanity. In the dialogue Socrates has asked that the hypothesis of "the first discourse" be read over to him. He then puts a question.

"What do you mean, Zeno? Is your argument that the existence of the Many necessarily involves like and unlike, and that this is impossible, since neither the like can be unlike, nor the unlike like. Is that your position?"

"Just that," said Zeno.

"And if the unlike cannot be like, or the like unlike, then neither can the Many exist, for that would imply an impossibility. Is it your purpose to disprove the existence of the Many? And is each of your treatises intended to furnish a separate proof of this, there being as many proofs in all as you have composed arguments, of the non-existence of the Many?"

"No," said Zeno. "You have misunderstood the general drift of the treatise."

After some further talk, Zeno unequivocally sets Socrates right.

"The truth is that these writings of mine were meant to protect the arguments of Parmenides against those who ridicule him, and urge the numerous fantastic and contradictory results which are supposed to follow from the assertion of the

One. My answer is addressed to the partisans of the Many, and intended to show that the greater or more ridiculous consequences follow from their hypothesis of the existence of the Many, if pursued, than from the hypothesis of the existence of the One." Incidentally he had invented dialectics.

Zeno then confesses that a love of controversy induced him to write his paradoxical treatise in the days of his youth. The book was stolen, he says, so he had no choice than to publish his paradoxes. "The motive" in publishing, he assures the somewhat skeptical Socrates, "was not the ambition of an old man but the pugnacity of a young one."

Whatever may have been Zeno's purpose in inventing his paradoxes, he was partly responsible for the failure of the Greek mathematicians to proceed boldly to an arithmetic of infinite numbers, an arithmetical theory of the continuum of real numbers, an analysis of motion, and a usable theory of continuous change generally. Hence any serious work in physics remained permanently beyond their capacities. They had traversed half or more of the hard way when they halted. Zeno's paradoxes and their own lack of an efficient symbolism for representing numbers stopped them.

The paradoxes, which a less fanatically logical people would have ignored for a season in order to get on to the real problems of developing arithmetic—finite and infinite—and creating a mathematics adequate for the study of physics and astronomy, made the precise, finite-minded Greek mathematicians over-cautious. They would consolidate and perfect what they already had, and make of it a single flawless masterpiece like one of their white temples on a bare hilltop. They succeeded in their theory of proportion, which stands today as perfect of its kind as it was twenty-three centuries ago, but unused and empty. By the time their masterpiece was finished for the admiration of posterity the great surge of their inventiveness had subsided in classicism and exhaustion. With the exception of the unorthodox Archimedes (287-212 B.C.), who did not disdain to think about things as well as about ideas, the leading Greek mathematicians after Plato belonged to their own memorable past. Fortunately for the progress of science and the advancement of mathematics, Newton in the 1660's ignored Zeno's paradoxes, if indeed he ever heard of them, and boldly created the pure and applied mathematics of continuous change. His reasoning about the "infinitely small" and the "infinitely great" would have shocked a mathematical purist of Plato's time. But it gave him the differential and integral calculus, without which neither his own astronomy and mechanics nor that of his successors in the eighteenth century would have been possible. He knew that his calculus was marred by logical imperfections, but he did not devote the youth of his intellect to the pursuit of a sterile purity.

Interpretations of Zeno's paradoxes of motion have been as numerous as varied, and as inconclusive as the guesses at his purpose in inventing them. Here the record is not one of thwarted progress, at least for philosophy. As Bertrand Russell remarked (in his Lowell Lectures of 1914), "Zeno's arguments, in some form, have afforded grounds for almost all the theories of space and time and infinity which have been constructed from his day to our own." Russell then states his own conclusions. On the assumption that finite spaces and times consist of a finite number of points and instants, Zeno's arguments, Russell asserts, are valid. "We may therefore escape from his paradoxes either by maintaining that, though space and time do consist of points and instants, the number of them in any finite interval is infinite; or by denying that space and time consist of points and instants at all; or lastly, by denying the reality of space and time altogether. It would seem that Zeno himself, as a supporter of Parmenides, drew the last of these three possible deductions, at any rate in regard to time. In this a very large number of philosophers have followed him." To which Zeno might reply, as he did to Socrates, "No. You have misunderstood the general drift of the treatise." In any case other paradoxes have appeared in the arithmetic of the infinite since Russell disposed of Zeno's. Russell continues, ". . . the difficulties can also be met if infinite numbers are admissible. And on grounds which are independent of space and time, infinite numbers, and series in which no two terms are consecutive, must in any case be admitted"—and so also, it would seem from the progress of the arithmetic of the infinite since 1914, must logical paradoxes of a kind unimagined by Zeno.

In addition to affording grounds for "almost all theories of space and time and infinity" from Zeno to Russell, Zeno's paradoxes have proved most stimulating to twentieth-century logic, especially to that part of it which evolved from admitting infinite numbers into mathematics. But the long-sought road to finality as pointed out by Russell was straight and clear in 1914: "It follows that, if we are to solve the whole class of difficulties derivable from Zeno's by analogy, we must discover some tenable theory of infinite numbers. What, then, are the difficulties which, until the last thirty years, led philosophers to the belief that infinite numbers are impossible? The difficulties are of two kinds, of which the first may be called sham, while the others involve, for their solution, a certain amount of new and not altogether easy thinking. The sham difficulties are those suggested by the etymology, and those suggested by confusion of the mathematical infinite with what philosophers impertinently call the 'true' infinite."

To which it may be added that the mathematical logicians —who certainly are a species, though perhaps a lowly one, of philosopher—since 1914 found it necessary to do much "not altogether easy thinking" about "the theory of infinite numbers" in the hope of making it "tenable." In the course of their thinking they evolved several new logical paradoxes, which may turn out to be sham, but which suggest nevertheless that there are more pitfalls and open wells in deductive reasoning than Thales or even Plato ever dreamed of. The new paradoxes now seem like natural consequences of the revolution in mathematical logic started in 1902 by Russell himself. Some of these will be noticed in the proper place.

Zeno's apparently inextinguishable paradoxes have been displayed here merely to illuminate the frozen peak of all philosophies of number, finite and infinite, the theory of Ideal Numbers as Plato visioned it in his maturest years. We shall endeavor to catch a glimpse of the unchanging reality which he described, after seeing what kind of man he was.

CHAPTER - - - - - - - - - - - - - - - 18

Politics and Geometry

"THERE are not in the world at any time more than a dozen persons who read and understand Plato: never enough to pay for an edition of his works; yet to every generation these come duly down, for the sake of those few persons, as if God brought them to hand."

This according to the Concord transcendentalist Ralph Waldo Emerson. He might have added that no two of the dozen at any one time are in complete agreement on what they read and understand. It does not necessarily follow that any of them have misread Plato. Where many consistent interpretations may be read out of, or into, some abstract general doctrine, it is not surprising that equally intelligent readers do not always agree on Plato's meaning. Fortunately for our purpose here, Plato repeatedly and emphatically said what arithmetic and geometry signified to him and implied for his philosophy. This being the only part of his system that concerns us, we may be reasonably sure that we understand what he meant.

Plato is usually and justly regarded as a pupil and disciple of Socrates. But he also had an older teacher who influenced his thought perhaps even more profoundly and whose disciple he might honestly have claimed to be. With the exception of the master himself Plato is the greatest of the Pythagoreans. He is much more than that, of course, because he is himself; but the important thing for us is that in him the Pythagorean numerology was perfected. After Plato the rest was slavish imitation or fantastically debased elaboration, even of the absurdities. And no man has made out a more plausible case for the existence of numbers and mathematical truths apart from the human mind than did Plato. Accepting the Pythagorean philosophy he codified and amplified it, and in his Ideal Numbers attempted to provide a rational basis for the "Everything is number" of his mystical predecessor.

Dismayed by the crudity of the master's generalization, some of the ancient commentators attempted to refine it and so clear Pythagoras of the charge of talking nonsense. For what he actually said and meant they substituted "Everything is represented by numbers." In support of their emendation they produced a letter signed with Theano's name. But the letter was easily shown to be a clumsy forgery. Theano had not betrayed her adored husband. As for Plato he sought no base subterfuge to escape the all but insurmountable obstacles presented by that crucial word "is" in the Pythagorean doctrine. "Is" belongs to the verb "to be"; and to prove incidentally, as one of the major problems of his entire system, that everything "is" Number, Plato imagined his theory of "being" as opposed to "becoming." It is sufficient for the moment to state that the outcome in mathematical realism continues to satisfy those mathematicians who believe that numbers were discovered rather than invented, and that "mathematical reality lies outside us."

Before following Plato into the nebulosities of Ideal Numbers we shall see what kind of man he was and what he thought of mathematics. As with other titans of the past his life is encrusted with legends. Some are obviously unfounded in fact, while others, though not plainly ridiculous, sorely distress his less discriminating admirers. Plato is a difficult man to write about objectively without giving unintended offense to at least a few. Incidents in his life that bring only an amused smile to the lips of the profane cause the bitterest anguish in the bosoms of the devout. Why not ignore them? Because not even a philosopher loses anything of his human integrity for being represented as a human being—if he happens to be as human as Plato was. He himself relates in his letters how his philosophy did not always work out in practice as it did in theory.

The official genealogy begins miraculously enough. On his father's side Plato (born either in the island of Aegina or in Athens, 427 or 428; died in Athens 347 B.C.) traced his lineage to the sea god Poseidon. When the supremacy of his intellect was recognized by his fellow Athenians, misguided enthusiasts provided him with a more direct celestial descent. This made him a half-brother to Pythagoras: his father was Apollo and he was born of a virgin. The only interest of these amusing fables today is what they tell us of the reverence in which Plato was held by his contemporaries and immediate successors. When the ancients could express their veneration for one of their great teachers in no more reasonable way they endowed him with a superhuman ancestry.

Whatever spiritual father Plato might have claimed, he was a son of Ariston, a descendant of the last King of Athens. Through his mother Perictione, he was in the direct line (sixth generation) from that same Solon who matched wits with Thales. He was thus doubly an Athenian patrician of the noblest class. From one of his distinguished birth and high opportunities great things were expected almost as a matter of necessity. The only question was what way of life he should follow. Politics was suggested—the Athenian government was in bad shape after the Peloponnesian war. As usually happens with young men of genius Plato made his own decision at the opportune moment. He chose philosophy and he chose it deliberately.

Little is definitely known of Plato's early years. His very name, Aristocles, sounds strange to us, "Plato," according to some authorities, being only a nickname meaning "broad." The most acceptable theory seems to be that "Plato" was bestowed on Aristocles by his instructor in wrestling, and referred to his shoulders. No anaemic student poring over musty old scrolls, young Plato submitted to a thorough drill in athletics as befitted a youth of his station, and is reputed to have won a wrestling match at the Olympic Games. He also amused himself by writing a great deal of lyric and dramatic poetry and composing an epic. The last was hastily destroyed when the young poet chanced to read Homer. The day before one of his dramas was to be performed in public, he more or less by accident happened on Socrates (469-399 B.C.) discoursing on philosophy. Here was what the restless young man had been subconsciously yearning for. He had already mastered rhetoric from the Heraclitean philosopher Cratylus, and had been stimulated by its literary and logical possibilities. But the discourse of Socrates was of a different order. It was philosophy and the real thing. Believing that he had at last found his vocation, Plato renounced literature and dedicated himself to philosophy with politics as a second interest. He burned all his poems and at the age of twenty attached himself as a disciple to Socrates. To judge by the poetic imagination displayed in his philosophical writings, Plato was not as bad a poet as he thought.

For eight years the young aristocrat frequented the society of the plebeian philosopher. Though tolerated by the regular followers of Socrates, Plato was not altogether welcome. He became less so when he began adulterating the Socratic philosophy with mystical refinements of his own. Even Socrates betrayed occasional irritation at what he considered his young admirer's callow philosophizing. Could he have lived to see himself in the "Socratic dialogues" of his most distinguished pupil he might have been disturbed indeed, for if there was one thing which Socrates, according to Xenophon, was not, it was a Pythagorean. He even might have been shocked to find his few undoctored teachings appearing as literature.

In perpetuating the Socratic dialogues Plato violated both his teacher's pedagogical theories and his own. Like his teacher, Plato professed to believe that the one effective way of imparting knowledge and fostering wisdom is by the spoken word. Socrates taught by talking, inventing for the purpose the famous "Socratic method." By skillful questioning he drew out of the pupil's mind what he supposed was already there, or what he himself had adroitly inserted. As for preserving any of his innumerable cross-examinations for the edification of posterity, Socrates was either too modest or too indolent to undertake the labor. Actually there is no evidence that he ever wrote a line, and but for the practical Xenophon and the poetic Plato, who became his teacher's posthumous amanuensis, we should know very little of the Socratic doctrine. Perhaps we know even less than we should have known if Plato had never met Socrates; for the report of what Socrates is alleged to have said was not written down till many years after his death.

In all the annals of education there surely can be no more strangely assorted master and pupil than the wise old Socrates and the brash young Plato. With no pretensions to social distinction of any kind, and fascinatingly ugly, Socrates was the plainest of the plain with no suspicion of mystery or pomposity about him. Dignity was foreign to him, for he knew that dignity may be the cloak of fools. Though he preached against some of the idols of "the mob," Socrates sensed how democracy might be practiced and how it should be encouraged. His lecture halls often were the street corners or wherever he happened to be when he felt like provoking an argument, and his auditors—to the shocked disgust of Plato's respectable friends—were as likely as not to be the ragtag and bobtail of the market place. The conservative citizens were unable to distinguish between the teacher and the irresponsible, loafing youths, rich and poor, who hung about arguing with him for hours on end. At last, with the restoration of the Athenian democracy in 399 B.C., he was summarily charged with contemning the officially recognized gods, diverting their due worship to divinities of his own invention, and generally corrupting the morals of the young—by teaching them that "Virtue is knowledge, vice is ignorance," and inciting them to use the brains with which they had been endowed by "whatever gods may be."

At the disorderly trial Plato tried to plead for his teacher but was shouted down by the rowdy judges. Condemned,

Socrates was granted a legal respite of some days to prepare for death. Plato sought to purchase his teacher's life but Socrates declined to be a party to the transaction. The jail doors were left open so that he might walk out and leave Athens, but he stayed. On the final day he and his friends conversed as usual till sunset, when the jailer brought the hemlock and Socrates drank. His last words were those of the plain man rather than the philosopher: "Crito, we owe Asclepius a cock. Will you remember to pay the debt?"

Plato, as he relates in the *Phaedo*, was prevented from being present at his master's death. But with the assistance of those who were, he restored the talk of that last day between Socrates and his friends. The topic was the immortality of the soul. The *Phaedo* is of particular interest to mathematicians because Plato made it the occasion for one of the most persuasive arguments in all of his writings for the extra-human existence of mathematical concepts. This will be noted later; for the present we remark a truly Pythagorean antithesis which could have been invented only by the narrator: "The end of life is death; the end of death is life."

Partly for political reasons, partly because the condemnation of Socrates had made Athens odious to him, Plato decided to travel after the death of his teacher. Always serious-minded he planned his journeys with but one gain in view: Knowledge and yet more knowledge. For twelve years he continued this roving education, adding to what he had learned from Socrates whatever lore of the Pythagoreans he could discover, and piecing together fragments of the philosophies of all schools then extant into the first rough pattern of his own. At Megara he mastered the tactics of disputation and deductive reasoning from the Eleatic philosopher Euclid.

(This Euclid is not to be confused with the geometer of the same name.) From Megara he crossed over to Cyrene in northern Africa, where Theodorus initiated him into the mysteries of irrational numbers. He is said by some ancient writers to have gone on (with Eudoxus?) to Egypt, where he received instruction in astronomy. As an aristocratic Athenian he ran a considerable risk in venturing into territory under Persian domination. To avoid unwelcome attentions he passed himself off as an oil merchant and succeeded in traversing the whole kingdom of Artaxerxes Mnenon in safety. Returning to Magna Graecia, he proceeded to Tarentum, where he studied intensively under Archytas, Timaeus, and other prominent Pythagoreans. Some of these men, particularly Archytas, were influential statesmen, a fact which probably saved Plato's life at a later stage of his career. We have already noted that Archytas is said to have presented Plato with a copy of the Pythagorean bible of Philolaus.

These travels were punctuated by visits to Athens. At the age of forty, being then exactly halfway through his earthly journey, Plato decided that his long preparatory education was complete. From Euclid of Megara he had learned the beginnings of dialectics; from Cratylus, rhetoric and natural philosophy; from Theodorus and others of the Cyrenaic school, mathematics and astronomy; possibly from the Egyptians, or from others familiar with their science, more astronomy; from Socrates, ethics, morals, and political theory; and last, from the Pythagoreans, everything. He was now ripe for the creation of an all-inclusive philosophy of his own. He would wander no more.

On his return to Athens his aristocratic admirers presented him with a plot of ground in a grove adjacent to the gymnasium. There he established his Academy. He never married. A modest house and a small garden provided him with all the space he needed for philosophical expansion and instruction. He took pupils but declined fees, although he was not averse to accepting substantial gifts. Knowledge, he said, is above all price and therefore should be freely bestowed. In welcome contrast to "the people's university" which Socrates had conducted, the Academy enjoyed the financial approval of all the best people of Athens, with the exception of Plato's prosperous competitors among the flashy and popular sophist philosophers. Amused by the contrast between the Socratic and Platonic schools, the comic poets unmercifully satirized the effeminate young gentlemen about town frequenting Plato's select Academy. Ignoring competitors and critics alike, Plato and his serious pupils went soberly about their business of constructing what is perhaps the most comprehensive system of philosophy the world has yet seen.

The Academy functioned for about nine hundred years. It was closed, never to reopen, in 529 A.D. by the celebrated if somewhat bigoted Christian emperor and legislator Justinian.

No philosopher has had a fairer opportunity than Plato to put some of his teachings into practice. The theory of government seems to have fascinated him as it had the Pythagoreans. His confidence in his own prescription for the ideal State, as formulated in his utopian *Republic*, was so unquestioning that he rather rashly welcomed an invitation to apply it to a singularly corrupt tyranny. In passing, the curious blend of idealism and ruthlessness proposed in the *Republic* as a permanent cure for all the ills of human society makes extremely interesting reading today. Such details as the communization of women, the state ownership of children, eugenics, the almost servile respect for the military caste, the arbitrary powers of the police, the control of science and religion in the interests of the state, the abolition of private property, the restriction of education to the few, propaganda instead of education for the masses, and the flat declaration that "in order to obtain the heritage, smaller nations must be trampled underfoot"—all have a familiar ring. Less common is the doctrine that states will know respite from their evils only when kings are philosophers and philosophers are kings. It was this millennial reformation which Plato attempted to impose on Dionysius the Elder, tyrant of Syracuse. Accounts of Plato's Sicilian adventures differ in particulars but agree in essentials.

Dionysius had a young brother-in-law, Dion, whose education was sadly in need of repairs when the tyrant politely invited Plato to visit Sicily. He was to make a general survey of conditions in the island, view the splendors of Mount Aetna, and put up for a while at the court in Syracuse. The tyrant's motive was not to honor a popular philosopher but to see what, if anything, could be done about Dion. It seemed to Dionysius that the young man was almost too able for his own well-being. The easy pleasures of a licentious court, which the tyrant himself relished exceedingly, had done Dion's constitution no noticeable good. Would the renowned philosopher care to take the young man in hand and try what he could do? Plato eagerly accepted. Here was a virgin mind, so to speak, to be trained from the beginning in the fundamental principles of sound government. Dion some day might rule Syracuse. Then, if all had gone as Plato planned it should, a philosopher would at last be king.

It was a case of love and mutual understanding at first sight between Plato and his somewhat debauched pupil. Dion appears to have had a first-rate mind, and the opportunity to use it on something more refractory than a courtesan gave him a rarer pleasure than any he had yet enjoyed. Hearing of virtue for perhaps the first time in his life, and following Plato's not altogether clear demonstration of the Socratic proposition that it is equivalent to knowledge, Dion was catastrophically converted to philosophy. His turn from an evil to a good way of living was as sudden and as lasting as a religious conversion. Ablaze with enthusiasm for the pure delights of knowledge and the profits to be derived from virtue, Dion burned to make a convert of his own. The tyrant himself was his chosen subject. But Dionysius was not yet eager to be virtuous or even literate. Recognizing that the case was too delicate for an amateur like himself, Dion called Plato into consultation. After many excuses Dionysius weakened and granted the philosopher an audience.

Plato's easy conquest of Dion had made him overconfident. Carried away by his zeal for the good life he proved to his own satisfaction that the injustices and brutalities of tyranny can give no pleasure but only pain to the tyrant. Dionysius was following the argument as closely as Plato could have wished. Unfortunately for both of them the teacher neglected to follow the play of emotions on the pupil's face. When it was too late Plato realized that he had gone much too fast. In a thundering rage Dionysius began shouting that he was not to be lectured and insulted by any pedagogue. Who was Plato to tell the King of Syracuse how he should govern? If this rubbish was the philosophy Dion was always prating about, he should have no more of it. The philosopher might go as soon as he liked—if he could. Plato got the point. Dionysius did not intend that he should continue living in Syracuse, though he was determined that the man who had made him feel like an ignorant bully should remain there indefinitely.

By great good fortune the ship which had brought Pollis, the Spartan envoy, to Syracuse was in the harbor at the moment and was just about to set sail for home. With Dion's assistance Plato got aboard in time to avoid assassination but not before Dionysius had spoken a word in private to Pollis. Between them it was arranged that Plato should be knocked on the head en voyage and his body pitched overboard or, if that proved inadvisable, that he be sold into slavery. Having no stomach for a murder that the Athenians were certain sooner or later to hear about, the diplomatic Pollis chose the second alternative. On reaching the island of Aegina—where some say Plato was born—he sold the broad-shouldered philosopher for a galley slave. This was a particularly dirty trick on the part of Pollis because Aegina and Athens were at war at the moment. He might at least have waited till the ship touched at a neutral port.

But all turned out well in the end. Plato was recognized by a fellow lover of wisdom. Anniceris of Cyrene redeemed the philosopher for about half a talent and waved him on, a free man, to Athens. When Plato's friends tried to repay Anniceris he gracefully declined to let them have the sole honor of serving philosophy.

On hearing of Plato's safe arrival in Athens, Dionysius realized that he had made a public fool of himself. By special envoy he sent the philosopher a suspiciously frank apology, begging him to return to Syracuse and give both Dion and himself further lectures on political science. Plato curtly replied that philosophy left him "no time to think of Dionysius."

With Plato no longer at his side to cheer and guide him, Dion grieved and intrigued. He worked assiduously on the tyrant's probable successor, trying to make a philosopher of him. When the tyrant died, and the rather dissolute Dionysius the Younger became the ruler of Syracuse, Dion felt it was time to summon Plato. It was now or never for the establishment of a government on sound metaphysical principles. The new tyrant himself was not wholly averse to the project. He had become vulgarly curious about Plato from all the wonders Dion never tired retailing of his absent teacher. He now declared that he too would be a philosopher. Thus, the young tyrant imagined, he might rectify his father's blunder and prove to those supercilious Athenians that he, at least, was no barbarian.

Sicily at the time was a favorite resort of the Pythagoreans, all of whom were devoted admirers of their most promising convert. The new tyrant had little difficulty in persuading these spotless souls to join him, his wife, and Dion in urging Plato to return and educate him. It is said that Dionysius' final inducement was a pledge to abandon his own form of government in favor of Plato's. Naturally he would need the philosopher in person to instruct him in the more abstruse details. Plato accepted and stepped into a mesh of political intrigue that was a masterpiece even for the tricky Sicilians. He was unaware that Dion's enemies had prevailed upon Dionysius to recall one Philistus from well-merited exile to oppose any adherents the philosopher might acquire. Everything that Plato disdained to be, Philistus was—an expert schemer, a passionate believer in tyranny as the ideal form of government, and a practical politician of the meanest stripe.

When his ship reached Syracuse, Plato was met at the pier by Dionysius himself. The elastic ruler had unbent so far that he was now leaning over backward to show his respect for the magnanimous pedagogue. He had driven down in his state chariot, himself the charioteer. He insisted on Plato getting in and being transported to the palace. Cheering crowds blocked the streets; the philosopher felt that he had come home to his own. Dionysius ordered a public thanksgiving and a sumptuous sacrifice to the immortal gods for the great man's safe arrival. For his part the tyrant proclaimed his intention to live a virtuous life, master philosophy, and govern his people wisely. Naturally the dissolute courtiers followed their beloved tyrant's lead.

To show that he meant what he had said, Dionysius at once began taking private lessons in geometry from Plato. The unfortunate courtiers also developed an insatiable thirst for triangles, en masse. Soon it became impossible for the slave boys, serving the sudden horde of geometers with pure and refreshing water, to put down a foot anywhere without disarraying the square on the hypothenuse in some enthusiast's diagram on the freshly sanded floor. The whole court hummed like a beehive in midsummer with definitions, axioms, and fragmentary demonstrations as overheated converts of both sexes argued to convince one another that they knew the subtle difference between a rectangle and a rhombus.

The honeymoon of politics and geometry lasted all of five days. Dionysius publicly announced that his own wooing of quadrilaterals had done his soul no end of good. To certify that he was as good a man as he felt he received petitioners of all stations with a humility and consideration he had never before shown. Under Plato's philosophical supervision, laws incorporating substantial reforms were drawn up and all but enacted, when Philistus jerked the tyrant's sleeve. He reminded Dionysius that Syracuse was at war with Carthage and whispered that Dion, with the connivance of Plato, was negotiating with the enemy and was about to reform the government, not by geometry, but by force.

Dionysius made an instantaneous recovery from his geometry. Once more himself, he ordered the arrest of Dion and posted spies about Plato's quarters. The philosopher was now a virtual prisoner of state. Still ostensibly friendly, the tyrant took adequate precautions to nip the philosopher's growing popularity with the masses. Simultaneously the courtiers discovered that they had never cared much about triangles and that rhombuses were a bore. The floors were swept clean of the gritty sand, couples began to dance, and once more the slave boys passed about with wine jars instead of water bottles. Dissipation succeeded disputation.

Disconcerted by this sudden downfall of virtue and the consequent triumph of ribaldry, Dion's faction became seriously alarmed for his and Plato's bodily safety, especially when it was rumored that Dion was to be unobtrusively knifed. Exerting all their very considerable influence to the limit, Dion's friends prevailed upon the tyrant to commute the death sentence to exile. Dionysius submitted and banished Dion to Italy. Promising to recall both Dion and Plato when, if ever, there should be peace with Carthage, he shipped Plato back to Athens. There Dion presently joined his teacher to receive further instruction in the philosophy of government.

For the second time Plato had bested a tyrant of Syracuse. However impracticable anyone might think the austere politics of the *Republic*, it was not Plato but Dionysius who shone in the public eye as an egregious ass. To preserve some shreds of his dignity Dionysius issued a wholesale invitation to philosophers throughout the civilized world to honor Syracuse with their presence and elevate all Sicily with their wisdom. He promised them quarters at the court and whatever luxuries they might require to facilitate their meditations. They accepted in droves. Even the Cynics were unable to resist this unique opportunity to despise wealth in public, while nakedly unashamed hedonists or money worshippers, like Aristippus or Aeschines, stampeded to the wallow. Dionysius felt that but for one detail he had redeemed himself in the eyes of the civilized world.

The detail was Plato. Without him even the snarling Diogenes was hardly convincing. But Plato was not interested in heading a company of braying nonentities. It was now the tyrant's turn to enlist the good offices of all his relatives, male and female. Adding their entreaties to those of their contrite master, they united in begging Plato to let bygones be bygones and restore the one true philosophy to Syracuse. Pleading the infirmities of age—a social fib—Plato declined. Only when Archytas and other influential Pythagoreans implored him in the name of philosophy to return to Syracuse and quell the madhouse gabble with some sound reasoning did Plato relent. For the third time he sailed for Syracuse in the hope of taming a tyrant and civilizing Sicily. The chariot in which Dionysius escorted him to the palace was twice as splendid as on the preceding occasion.

This time the political horizon showed not a cloud. Philistus had earned the lasting hatred of all the common people of Sicily, and if Dion could have accompanied Plato he would have become the long-desired philosopher king.

Dionysius, too, was a changed man. Mild and gentle, almost melancholy, he listened with wistful patience to Plato's measured denunciation of the evils of tyranny, and promised to reform. He promised anything in fact except to keep his promises. Plato reminded him that Dion was to have returned with the restoration of peace. Hostilities had ceased long since, and Dion still was in exile. Yes, the tyrant sighed, it was all true, but he could do nothing about it; Dion refused to return. Plato realized immediately that he had been duped again. He remembered that the last time he saw Dion, that thoroughly regenerated young man was impatiently awaiting a signal from Dionysius to return to Syracuse and put the Platonic principles of government into instant operation. The most serious obstacle, Dion declared, had been removed. The tyrant would agree to a redistribution of property in which all should share equitably. Now, it was clear to Plato, if there was one thing to which Dionysius was constitutionally averse, it was communism. Their intimacy became a sour farce, maintained on both sides only to save face before the jealous horde of gossiping philosophers of all denominations swarming about the court. These disgruntled lovers of wisdom insinuated to Dionysius that Plato kept himself aloof from them because he was conceited enough to imagine that he could run the government better than its lawful head.

Mindful of his past blunders, Dionysius remained suavely

wary. When Plato could support his anomalous position no longer, he requested permission to return to Athens. With a show of courteous reluctance the tyrant assented. A ship of state was placed at the philosopher's disposal. Then, at the critical moment, Dionysius changed his mind. Before the ship could clear the harbor Dionysius signaled the captain to put about. Plato was taken off and placed under arrest. Dionysius was not going to have this unruly philosopher talking about him to those smug Athenians.

This move was too much for Archytas and Plato's other Pythagorean friends in Tarentum. Archytas sent Dionysius what amounted to a sharp ultimatum demanding Plato's immediate release. Here was no abstract proposition of political theory but a concrete threat of war. Dionysius understood. As a last face-saving gesture he ordered the most lavish dinner Syracuse had ever put together, burdened the departing guest of honor with a wealth of rich gifts which he did not particularly desire, and personally escorted him to the ship. As Plato stepped aboard Dionysius whispered a last request: "Think of me sometimes when you are not thinking of philosophy."

On the return voyage Plato paused at Elis to witness the Olympic Games. His presence excited more interest among the crowds than all the athletic contests some of them had come hundreds of miles to see. He was the hero of Greece.

The remainder of his life was passed quietly and without interruption in his garden and house with his pupils. He died in his eightieth year at a wedding feast. He was buried under the trees he had loved.

CHAPTER - - - - - - - - - - - - - - 19

"Another I"

WE WERE reminded that "many philosophers of high reputation from Plato onwards" have held the view "in one form or another" that "mathematical reality lies outside us." It will be interesting and instructive to see what led Plato to this remarkable conclusion. Even a cursory reading of Plato's works suffices to show that the elementary properties of numbers and the tactics of geometrical proof profoundly influenced his thought in the elaboration of his entire philosophy.

Before we consider his argument for the "reality" of mathematics, it will be well to summarize what he actually said about arithmetic and geometry, both on their own account and as aids to philosophic thought. We shall then be in a position to see why he formed so high an opinion of mathematics in its least useful aspects. What he had to say of mathematics was repeated, sometimes with understanding but more frequently with uncritical adulation, by the scholars of the Renaissance. In the Platonic revival of the fifteenth and sixteenth centuries some of the more rhetorical commentators on mathematics even sought to surpass Plato in their praises of "the science divine," and a few succeeded in classic passages of moving eloquence. These men will have their say later. Though not himself a mathematician of the stature of a Eudoxus or an Archimedes, Plato was almost obsequiously complimentary to pure mathematics. Some of the flattering things he said of arithmetic and geometry as disciplines for the philosophic soul and revealers of eternal truths may sound a trifle exaggerated today; but he said them so beautifully that it would be ungracious for any modern mathematician to quarrel with him. His appraisal of mathematics was what might have been anticipated from an aristocrat of the aristocrats and a philosopher whose primary passion was for morals and ethics.

Plato's major mathematical problem was twofold. The abstractness of the data of mathematics—its numbers, its points, its lines—suggested a specious argument for the existence of "entities" directly perceptible by the mind and independent of sensory experience. These extra-sensory entities predicated a suprahuman realm of eternally existing Forms or Ideas, in which the "truths" of mathematics "participate."

The first part of Plato's problem was to establish these Ideas beyond any rational doubt and from them to infer the phenomena of the sensory world. The second part was so intimately interconnected with the first that the solution of either would imply that of the other. The Heraclitean flux —"All things flow"—was repugnant to a mind fixed on an eternity in which there is neither change nor shadow of change. The world of the senses, as the Pythagoreans had insisted, is notoriously unstable and impermanent. If "things are not what they seem," what are they? Plato's answer was that the partial truths, the flawed beauties, and the imperfect goodnesses accessible to the senses or the intellect are mere "becomings," first approximations toward full participation in an absolute Truth, an absolute Beauty, and an absolute Good existing permanently in a realm of eternal Being. His central problem was to prove the existence of these absolutes, especially that of the Good; and it seemed to him, plausibly enough, that mathematics offered the one cogent analogy and the sole hope of success.

What Thales and Pythagoras began Plato finished. On the terrestrial level he strove to perfect the arithmetical synthesis of the universe first proposed by Pythagoras. All the scientific myths of his predecessors in philosophy, and much of the prescientific mythology of a more ancient wisdom, streamed through his mind, to issue in one great river of unified thought, that swept through the early centuries of the Christian era to the Middle Ages, and thence through the scientific renaissance of the sixteenth and seventeenth centuries, gathering volume from the imaginings of innumerable idealists and numerologists in its irresistible plunge from the past to the present.

On the celestial level at least two of the Platonic absolutes have survived in the beliefs of millions, the Good as the deity, and Truth as its eternally incorruptible self. The Beautiful as an absolute seems to have passed out of time. "Beauty lives only in the beholder's eye"; it is a matter of personal taste and judgment. But not so for Truth, especially as revealed by mathematics. To the modern Pythagoreans mathematical truth is the same partial projection of the absolute Truth that it was for Plato. On both the terrestrial and the celestial levels the apparent inevitability of the conclusions reached by mathematical reasoning was a primary source of all the

Platonic philosophy. Twice two might mean four to the senses, but to the soul its meanings were infinite.

The philosopher, Plato asserts, must be a (Pythagorean) arithmetician. He should contemplate number till its inner nature is perceived only by the mind, and this he must do for the well-being of the soul itself. For number is the most direct of all means for passing from "becoming" to "being," from change and decay to permanence and immortality. Indeed number exists primarily so that the soul may ascend from the transitory to the timeless and share in the everlasting. Geometry also withdraws the soul from becoming to being and conditions it for participation in the Good. The real object of both is knowledge; and that knowledge toward which arithmetic and geometry strive is not of perishable things, but of the eternal. Music too, if mathematicized and directed to the Beautiful and the Good, will draw the soul toward Truth and foster the spirit of wisdom.

On a more mundane level arithmetic is the primary kind of knowledge in which the noblest natures should be expert. It is said to be essential to complete manhood because of its singularly elevating effect on the human mind. Indeed all mathematics is indispensable for heroes, demigods, gods, and any others who may aspire to the sublimest knowledge. Particularly is mathematical knowledge necessary to the gods, for in mathematics there is an element of fate which not even the deity may defy.

As for mathematics itself, it soars above the deceits of the senses to eternal freedom in the realm of absolute realities. As Thales may have dimly imagined in his abstraction of the data of the senses, Plato emphasizes that geometers are not concerned with the visible lines of their diagrams, but are contemplating "the absolute square, the absolute diameter," and so on—the "things in themselves" which can be "seen" only with the mind. Though abstraction may proceed from sensory experience the truth which mathematics discovers is in no degree sensual or variable, like opinion, but is ideal and absolute, in short, knowledge. The reason, or even the soul, has no part in the creation of mathematical truth, but merely is aware of its existence when properly disciplined. This is the point where the disciples of Pythagoras, Plato among them, diverge from a majority of twentieth-century mathematicians.

One of the philosopher's most curious arguments for the independent existence of mathematical truths is that the human body has no sense organ adapted to register them. Being beyond the range of sight, hearing, smell, taste, and touch, and yet being apprehended by the mind or the soul, these truths must exist independently of the senses. Their incontestable existence resolves the perpetual conflict between the senses and the intellect, between opinion and knowledge, between appearance and reality, and is alone sufficient evidence of a suprahuman realm of invariable Being.

The sciences likewise testify to a permanence beyond all change, but only in so far as they present their conclusions through arithmetic and geometry. It follows that the relative reliabilities of several sciences can be justly estimated by the amounts of mathematics they contain. For "the deity ever geometrizes," and what is not in harmony with geometry can be only an illusion of the absolute reality excogitated by the deity.

The much-quoted aphorism about the geometrizing deity looks like a slip of the pen or a temporary absence of mind on Plato's part. Actually it does not occur in his writings. It is merely attributed to him. Certainly it is contrary to the true Pythagorean faith which Plato held always before him and which he stabilized in his own Ideal Numbers. An amended version, proposed in the nineteenth century by another great Pythagorean (C. G. J. Jacobi, 1804-1851), is closer to Plato's philosophy, "the deity ever arithmetizes." As further amended by one of the greatest arithmeticians in history (J. W. R. Dedekind, 1831-1916), who was in no sense a Pythagorean, this becomes the entirely human variant of the original, "man ever arithmetizes." Between the first and the last lie about twenty-three centuries of discordant philosophies all appealing to mathematics to certify their ultimate validity.

If Plato could glorify useless mathematics with the tongue of an inspired angel, he could also denigrate useful mathematics with all the scorn and contempt of an angry man annoyed beyond endurance by what appeared to him as the ugliness and triviality of mere living. His philosophic calm vanishes before the spectacle of divine mathematics in the lowly service of humanity. He admits that ignorance of the elementary applications of arithmetic and geometry is ridiculous and reprehensible. Those who do not know these simple things, he says, are more like swine than men. But there he stops, harshly deprecating utility as a motive for the study of arithmetic and geometry. All useful arts, he asserts, are ignoble and inherently mean. Those who will see in his own mathematical imaginings "only idle fables because no material profit is to be derived from them" may merit his sarcastic abuse. But what of the astronomers he so soundly trounces for presuming to check the motions of the invisible planets of their calculations against observation of the planets in the sky? Or what of the physicists who venture to ascertain the facts of acoustics by plucking strings, who also are denounced as betrayers of Truth and traitors to their own higher natures? Even the professed mathematicians come in for a share of the philosopher's scorn. They are alleged to confuse the necessities of daily life with those of geometry, whereas the real object of geometry is knowledge of the absolute Truth. Geometers may have been like that in Plato's Academy. But no geometer since then has been so stupid as to be misled by an atrocious pun on the meanings of "necessity." Nor, for that matter, was either Thales or Pythagoras so obtuse. As for the censure of observational astronomers and experimental physicists, it recoiled upon itself in several hundred years of needless sterility in science for lack of direct contact with nature.

But all these strictures on the useful mathematics and the empirical science which Plato despised can be—and have been—set aside as merely the evidence of a justifiable irritation with those who failed to appreciate his larger purpose. In his own time Plato was to the average scientist what a theoretical physicist is to a dabbler in haphazard experiment today. He was not seeking interesting or spectacular happenings. What he sought was a simple generalization to coordinate all phenomena and a method that would infallibly reveal whatever enduring reality may be concealed in any testimony of the senses. His Ideal Numbers may have given him the required generalization; his dialectic, he believed, supplied him with the method. For each of these an idealized theory of mathematical procedures and mathematical truths appeared to be a necessary preliminary. Today his purpose animates the dream of the modern Pythagoreans, who substitute the principles of epistemology for Plato's Ideal Numbers and mathematical analysis for his dialectic. Though transposed to a higher key the ancient melody is easily recognizable and the lyric is the same: everything is number; observation and experiment are superfluous and misleading.

Any isolated remark of Plato's concerning mathematics may give a totally erroneous impression of what he actually thought of the subject. For example, it is said—on doubtful authority—that over the entrance to his Academy he posted the ban: "Let no one ignorant of geometry enter my doors." This was not a testimonial to his esteem for geometry in itself. Whether or not the story is a myth, Plato imposed the entrance requirement of geometry so that the tough-minded students of his own dialectic might have some skill in the rudiments of logical reasoning.

Philosophy in the Academy was no idle pastime for dilettant quibblers, but a very serious business indeed for mentally mature young men. Roughly, a good deal of it corresponded to a graduate course in mathematical logic today. It was the metamathematics of its period—a critical examination of the grounds of beliefs, hypotheses, postulates, and modes of reasoning, mathematical and other.

If some of what was debated in the Academy is no longer of interest to mathematicians or modern logicians, Plato should at least be credited with having induced mathematicians to inquire what, if anything, it may be that they are talking about. He and his pupils developed one of the major creeds concerning the nature of mathematical truths; and if some mathematicians today find the Platonic reality of such truths childish if not downright absurd, some others, fully as competent, accept the Platonic creed as reasonable and satisfying. So whatever anyone may think of Plato's mathematical philosophy, he can neither establish nor dispose of it by citing eminent authorities in mathematics. For as Socrates might have pointed out, the Platonic reality of mathematics is a matter of opinion, not of knowledge, and debates on its validity are battles of words about nothing.

But the futility of a debate does not imply that the topic of debate is of no importance for the debaters. The mathematician who believes what Plato taught about mathematics may find nothing preposterous or disturbing in the abandonment of scientific observation and experiment in favor of the unaided reason. More probably he will favor any mathematics but the purest with his contemptuous scorn, especially if his entire life since childhood has been passed in the sheltered obscurity of a lecture room. His opponent may find much to upset him, and if he be of a pessimistic turn, may even be moved to prophesy a recurrence of the Dark Ages. For belief in the Platonic reality of mathematics seems to be a fairly reliable touchstone to discriminate between the ancient or mediaeval scientific mind and the modern, or, as a modernist might say, between the unscientific and the scientific. Mathematicians as a rule are unscientific. This is not the opinion of mathematicians but of scientists. It can be checked by polling men who make their livings at science.

Since most of Plato's mathematical creed is put into the mouths of persons in his dialogues, we may not know what either they or he actually believed about mathematics. Only once, for instance, does the Socrates of the dialogues speak out in his traditional character. Unlike Plato, Socrates had no very exalted opinion of mathematics, either as a revelation of Truth or as a training in reasoning. He granted that geometry is useful in measuring fields, and the like, and that was all. In the *Republic* Glaucon asks, "Surely you could not regard the skilled mathematician as a dialectician?" Socrates hastens to reply: "Certainly not. I have never known a mathematician who was capable of reasoning." That at any rate sounds like honest reporting. Although Socrates may have overstated his case somewhat, there are many today who would agree with him. But never Plato.

Mathematics, he declares, quickens the general mentality and is invaluable as a preliminary discipline for young men not sufficiently mature to begin the hard business of philosophy, dialectical argument, and Pythagorean science—numerology. The contribution of a mathematical training to the serious purposes of philosophy is direct and positive. It is the mathematical method rather than the truths of mathematics which is all-important. The future philosopher masters, through his geometrical exercises, the correct concept and function of definitions, strict deduction, the technique of analysis and the indirect method of proof (described here in an earlier chapter), both indispensable in dialectic, and the organization of thought. Such training is necessary for all who would find knowledge; it prepares the mind to seek and recognize ultimate realities as opposed to the evidence of the senses. Mathematics cannot of itself reveal ultimate reality or absolute truth; dialectic can. Opinion is of the senses and is concerned with "becoming"; knowledge is of the mind and relates to "being"; mathematics is a bridge between opinion and knowledge.

More subtle and incisive than mathematics, dialectic is any process which isolates new truths by analysis and argumentation. Wholly of the mind, it proceeds from Ideas, through Ideas, to Ideas. In a purely mathematical investigation the validity of the hypotheses is not questioned. Dialectic seeks—and finds—in the Ideas the realities validating the assumptions of mathematics. It justifies the "self-evidence" of the mathematician's axioms—described in Plato's day as "common notions," and for long thereafter as "self-evident truths"—and examines the basic hypotheses and fundamental processes of all methods of discovering truths, of which mathematics is only one.

If Plato were writing today he might claim that what he was really talking about was the metamathematics and the metalogic of the 1930's. Two quite simple illustrations of his appeal to mathematics to clarify a metaphysical argument occur in the *Meno* and the *Phaedo*. In the former it is asked whether virtue can be taught; in the latter, whether the soul is immortal. "As in geometry" hypotheses are assumed and their consequences analyzed, almost as if one were trying to prove a conjectural theorem. Both arguments will be summarized in the following chapter. The second is perhaps the clearest example of the kind of evidence that inspired Plato to invent his Eternal Ideas.

To Plato also can be traced the persistent dogma that mathematics should form the basis of a sound education. On his return to Athens after his travels in Italy and the East, he contrasted the lack of training of Greek boys in arithmetic and geometry with the thorough drill of Egyptian schoolboys in these useful subjects. But he did not dwell too long on base utility. Thus in the *Republic* he prescribes an intensive mathematical education for the guardians of his ideal city because, he asserts, all the arts and sciences necessarily involve number and calculation. While admitting that not many subjects are as difficult as mathematics—in particular arithmetic—for the average mind, he encourages the timid by assuring them that number when studied for its own sake is entrancing, and the more abstract arithmetic is the better it is for the soul. On a less elevated level, arithmetic and geometry are said to be indispensable in military tactics, enabling a commander to dispose his troops to the best advantage. Here, so far as arithmetic is concerned, Plato may have been visualizing the square, triangular, and oblong numbers of the Pythagoreans in battle array. Such formations were actually used. But always he ends on an idealistic note: the one truly worthy purpose of mathematical studies is to draw the soul toward Being.

Were he living today Plato would find himself at odds with those psychologists who claim to have shown statistically that there is little or no transfer of training from one subject to another, and that mathematics as a general mental discipline is not what our fathers believed it was. Whatever may eventually emerge as the fact in this somewhat acrimonious dispute over the value of mathematics in a general education, there can be no doubt that Plato's authority reinforced the Pythagorean insistence on the high value of a thorough training in the rudiments of mathematics, and aided in retaining arithmetic and geometry in the school curriculum for over two thousand years.

When asked what a friend is, Pythagoras answered "another I." Pure mathematics never had a better friend than Plato, nor Plato a better friend than pure mathematics.

CHAPTER - - - - - - - - - - - - - 20

Number Deified

WE PASS on to a brief indication of some of the evidence which Plato offered in support of his "mathematical realism." First there is a detail of technical language. Since his Ideas are the enduring "realities" for Plato, his system is a species of "realism," despite its concern with ideal "entities"—the Ideas—beyond direct experience by the senses.

A cornerstone of Platonic mathematical realism is the doctrine of anamnesis or recollection. It is graphically presented in the dialogue *Meno*. Socrates and Meno have been arguing about the possibility of teaching virtue. Socrates undertakes to prove that "there is no teaching"—in the sense of one mind conveying or transmitting knowledge to another—"but only recollection." He asks Meno for one of his "numerous attendants" to serve as the vile body in his purposed demonstration. A presumably ignorant but intelligent slave boy who understands Greek and who "was born in the house," is just what Socrates wants. "Attend now," he says to Meno, "to the questions which I ask him, and observe whether he learns of me or only remembers."

By ingenious leading questions and a simple geometrical diagram Socrates induces the boy to "remember" some prenatal mathematics. He is led, for example, to reason out $3 \times 3 = 9$, $2 \times 4 = 8$, and to read off from the diagram that 8 is not the square of 3. His inability to point out the square root of 2 on the diagram finally stops him. But he has "remembered" that the square of twice a number is not twice the square of that number, and actually recognizes the square root of 8. After some further coaxing Socrates draws Meno's conclusions for him. They are momentous.

According to Socrates the experiment has shown that the slave did not know what was in him waiting to be drawn out. The boy's ability to give correct answers to the questions proves that the mathematical truths dormant in his mind "are just waking up in him," under the questioning, "as in a dream." Further, "the knowledge which he now has" he must "either have acquired or always possessed." But as the boy was innocent of any schooling in mathematics, the second of these alternatives must be admitted.

Socrates appears to believe that he has established his thesis. Mathematical knowledge is of the eternal. Our souls knew it before we were born, forgot it on entering this life, but may recall it by concentrated effort on being properly stimulated. In particular, mathematics is not created by the mind but is only "remembered." The grand conclusion follows: "and if the truth of all things always existed in the soul, then the soul is immortal. Wherefore be of good cheer, and try to recollect what you do not know, or rather do not remember." To which Meno—not depicted in the dialogue as a skeptic—replies, "I feel somehow that I like what you are saying." Not to be outdone in politeness, Socrates caps this with a compliment to himself, "And I, Meno, like what I am saying."

A famous exposition of the doctrine of anamnesis, on purely intuitive grounds, is Wordsworth's in his *Ode on the Intimations of Immortality.* Like most poets who have been enthralled in their youth by Platonic realism Wordsworth did not believe, as Socrates (Plato?) evidently did, that he had given a logical or scientific demonstration of the immortality of the soul. Realism in the Platonic sense is an affair of the emotions, not of the reason. Mystics, mathematical and other, find it acceptable. One meaning of mysticism is immediate knowledge of the real by direct intuition without the mediation of sense or reason. A true mystic has no need of any such demonstration as that of Socrates. For him it is not superfluous but irrelevant and meaningless.

Having proved the immortality of the soul, Socrates ascends rapidly to the immortality of virtue. It is unnecessary for us to follow him. All of his—or Plato's—attempted proofs for propositions of this kind are fundamentally the same. To a mind eager to be convinced the most convincing of all Plato's efforts to establish "the objective reality of universals"—like truth, virtue, love, man, knowledge, and so on—are those concerning the common notions of arithmetic and geometry. And so it has been since Socrates convinced Meno. Plato was neither the first nor the last to seek mathematical permanence in the elusive memories of things past. For century after century thinking man's urge to find some abiding refuge in the eternal flux has swept him time after time to the rock —or the reed—of a supposedly eternal mathematics.

The core of all Platonic realism is the mystical doctrine of Ideas or Forms. According to some experts Plato's writings contain at least two theories of Ideas. Mathematicians who agree with Hardy (quoted in the first chapter) that "mathematical reality lies outside us" need not concern themselves which theory of Ideas they cite in support of their belief. One, or a fusion of all, will provide them with an abundance of mutually consistent arguments. Nor need those who wonder with Kasner how human beings ever came to believe in the Platonic reality of mathematics be troubled by any scholarly doubts about which version of his theory, if any, Plato himself finally believed. Any single specimen of all the varied arguments will suffice to show what induced realistic mathematicians to vision the "objective reality" of mathematics. As it is one of the simplest we shall consider that in the *Phaedo* concerning "real" equality.

The senses never report any two things as exactly equal; refined measurement always reveals a further discrepancy not detected by cruder observations. So far as sensory evidence can tell us anything there is no limit to the sequence of refinements. Yet, although exact equality is beyond the reach of the senses, the mind has no difficulty in conceiving equality with absolute exactness. If this "real" equality is forever inaccessible to sensory observation, where is it and what is it?

A logical positivist of the extreme school might say that this question is meaningless. An operationist would claim that any proposition, for example, "the lengths of these two rods are equal," has meaning only if the proposition includes the means of its verification. "Absolute equality exists" is a proposition which fails to meet this test. It is a specimen of metaphysical statements that are scientifically meaningless. Such statements are not necessarily unimportant. Many of them have determined human conduct and some have occasioned long and sanguinary wars. They merely have no significance for science. Are they relevant for pure mathematics? A mathematical realist has no doubts on the matter. For him absolute, perfect equality exists. It is the Idea "Equality" in which the equality of the senses participates but never fully attains.

Equality as an Idea is eternally the same. Being invariable, this real Equality is the only possible object of knowledge in all questions of equality. For if the measured lengths of two metal rods, say, "equal" in length by the micrometer, are continually changing with the temperature, and the micrometer itself is palpitating in an extremely erratic manner, who really knows that the rods are actually equal in length? And what does it mean to say that they are? The current scientific answer that these questions can have no significance because all empirical measurement is statistical in character merely sets the enquiry back a step. The senses, as in scientific experiment, generate opinion; the reason, as in mathematics, generates knowledge. This appears to be how the mathematical realist, following Plato, distinguishes between opinion and knowledge. Elaborating his position the realist criticizes the interpretation of a typical experiment.

Experimenter A is of the opinion that the sixth decimal in the length of a rod measured against a common standard is 7 with a probable error of plus or minus 2. In experimenter B's opinion 5 plus or minus 1 is right. The mathematical realist insists that neither A nor B can know anything whatever about the "real" dimensions of the rod so long as either continues appealing to his instruments and his senses. Now the numbers corresponding to the lengths that A and B put into their formulas and equations, also any accompanying probable errors, are assumed to be invariable throughout the processes of mathematical deduction. While they are reasoning correctly, the realist maintains, A and B are in the realm of knowledge. When they cease reasoning abstractly and translate their deductions into experiments, they have lapsed back into the flux of mere opinion.

Such is approximately the realist's opinion, and he never ceases to marvel that experimentalists imagine they are increasing the stock of human knowledge. For him the only part of any science that can be rightly called knowledge is the mathematical. Hence science, like a split personality, is in perpetual conflict with itself while mathematics, moving wholly in the realm of reason, is self-consistent and sane.

The endless strife between Socratic Knowledge and opinion is reflected in science as a disagreement between theory and observation. That such disagreements do occur, and quite frequently, is undeniable. Which one may be at fault is not yet decided, but it has been suspected in some instances that both may be guilty. The only thing of which the realist is sure is that his mathematics is eternally true and therefore everlastingly right. In itself it is, as the realist defines truth; but the realist's mathematics does not therefore have any significance in the worlds of scientific experience and good sense.

Perhaps even more suggestively than the common notions of mathematics—equality, points, lines, and so on—its theorems supply the mathematical realist with innumerable confirmations of his faith. Instead of citing Plato (or Socrates) on this important detail we shall quote another renowned philosopher who was also a mathematician of high rank. We appeal to Descartes (1596-1650) because he typifies the great mathematician whose feeling for science is metaphysical and uncertain. As might be anticipated of such a mind, Descartes was an unquestioning believer in the objective reality of mathematical concepts. He beheld the vision of the Real (in his *Fifth Meditation*) in his Eternal Triangle.

"I imagine a triangle," he says, "although such a figure perhaps does not exist and never has existed anywhere in the world outside my thought. Nevertheless this figure has a certain nature or form or determinate essence which is immutable and eternal, and which I have not invented and which in no way depends on my mind. This is evident because I can demonstrate various properties of the triangle, for example that its three interior angles are together equal to two right angles, that the greatest angle is opposite the greatest side, and so on. Whether I wish to or not, I recognize very clearly and evidently that these properties are in the triangle although I have never thought about them before, and even if this is the first time I have imagined a triangle. Nevertheless, nobody can say I invented or imagined them."

The mysterious triangle whose properties Descartes imagines he has not imagined is the universal Triangle, that particular Platonic Idea in which all triangles recorded by the senses participate by virtue of their triangularity. To realists Descartes' argument is clear and convincing. Others, it is only fair to state, find it delightfully naive. Once more it is a matter for the emotions rather than the reason to judge.

So long as Plato "realized" the abstractions of mathematics, aesthetics, ethics, and morals in Ideas he seems to have felt reasonably sure of himself and his realism. But when less pleasant things insisted on their metaphysical rights and also became copies of corresponding Eternal Ideas he began to hesitate. Intermediate between such innocuous Ideas as Equality and the sublimest of all—Truth, Beauty and the Good—were the Ideas corresponding to commonplace but unobjectionable things such as plants and animals. Were these Ideas entirely clear? Though the question was intended in a different sense, Plato might have asked, addressing his query to the Good, "What is Man that thou art mindful of him?" The answer is simple.

Of the hundreds of millions of individuals of whom it can be asserted "this is a man," not one is Man. As Protagoras observed, you never see Man walking down the street; you see a man, and recognize him as such even if you do not know him. But ignoring this sophistry as Socrates did, the realist asserts that each individual man participates in the Idea "Man." The universally predicated general term "Man" denotes a certain Reality. This Reality can be apprehended by the reason but not by the senses; men, not "Man" are what the senses report. The Many who are men share in the One that is Man, and Man exists in the forever changeless realm of Ideas as an "objective reality."

Similarly, passing to intangibles, the realist may first imagine all the beautiful things in the universe spread out before him. None of them is wholly beautiful and few are alike. Yet they all share in a certain "essence" which the realist recognizes as beauty, and the concept Beauty, in which all beautiful things participate, comes unsought to his mind. And this Beauty he feels is an "objective reality"—more real, and in a different and permanent way, than any of the perishable objects he has recognized as beautiful. Sensing that Beauty is somehow good, and the Good somehow beautiful, and both somehow true, the realist then experiences a mystic revelation that the Beautiful, the True, the Good, as Ideas, somehow participate in one another. And since arithmetic and geometry participate in the True, therefore they also are beautiful and good.

With this agreeable conclusion the average mathematical realist is usually content to rest his case. But if he is as ruthlessly metaphysical as Plato he will continue to less pleasing Ideas. What of the Ideas "Nail-Paring," "Hair," "Dirt," "Filth," and the like? Are these too absolute and eternal with the others? As Ideas they are. When Parmenides put this disturbing question to Socrates, then a young man, the latter was revolted and denied such gross or unclean things the privilege of participating with the True, the Beautiful, and the Good in Being. For Being is the Idea in which every Idea participates fully, since otherwise it would not be an unchanging Idea at all, but a mere delusion of the senses and a transient "becoming." Parmenides assured Socrates that philosophical maturity would cure him of his youthful squeamishness about Reality, and indeed it did. His recovery restored him to humanity.

In implying that Plato believed in the objective reality of his Ideas we are on disputed ground. The weight of philosophical authority seems to favor this opinion, though some critics contend that Plato in his old age abandoned the objective reality of the Ideas, and cite passages from the *Parmenides* (and others of Plato's writings) in support of their contention. It is immaterial to mathematical realists which side is wrong; for they, certainly, adhere to the objective reality of mathematical Ideas whatever Plato may have believed at the last. Otherwise their realism makes no sense of any kind.

The Ideas, although objective realities, are not "things" as things are meant in common language—bricks, people, emotions—but are akin to thoughts. An Idea however is not a thought in any man's mind, nor yet in the Absolute's mind —if the Absolute (or the Good) has a mind or if, having a mind, the Absolute has thoughts. The Ideas are self-existent entities that can be thought about by a thinker. They are extra-spacial, extra-temporal; independent of all gods and of all men; eternal, unchanging, and perfect; uncreated by reason but apprehended by it; and "known" only through the reason —or the soul—not through the senses.

All this is evident to the mathematical realist. But subtle difficulties present themselves. The ancient conflicts of the One and the Many break out again in the distinctions between the Ideas and the world of the senses. Thus there are any conceivable number of triangles but only one Triangle, and an infinity of integers but only one Integer. The realm of Ideas is that of Absolute Reality, Absolute Being; the world of the senses is unreal and unstable, except in so far as the objects of sense participate in, or partake of, Ideas. An Idea is a One shared through a partial reality by a Many—if that Many have any reality at all. Though realism as an Idea is a One its obscurities, as developed by Plato in the *Parmenides*, are many. For example, what is the "real" status of such a proposition as "twice three is seven"? In what Idea is it participating? The solution is almost obvious—to a realist.

To complete his Absolute Being and round out his Absolute Reality, Plato invoked his creative imagination and easily surpassed the feat of the Pythagoreans when they invented the

Counter-Earth. He convinced himself that he had conceived an Absolute Not-Being. The "unreal" constituents of objects in the world of the senses, also the false propositions published from time to time by careless realists, such as "twice three is seven," participate in this Absolute Not-Being.

The eternal existence of this monster need not dismay us, for it is mainly through participation in Truth, Beauty, and Goodness that realism validates both mathematics and many an ancient creed that has survived to this day. But we should not forget that Zeno told Socrates the paradoxes of the One and the Many inspired him to the invention of his own. Until all of these, including the capital paradox of the One and the Many itself have been satisfactorily dissipated, mathematical realism (like mathematical analysis) may lack a consistent foundation. This however need not be a serious objection. It has yet to be shown that inconsistency is necessarily antipathetic to belief.

Having caught a fleeting glimpse of Ideas as "objectified concepts," we now ask how is an Idea to be revealed? Not by the senses, certainly. Not in its entirety by the reason, probably. But the reason can approximate to such recognition by the practice of dialectic.

Wholly divorced from the senses, dialectic operates exclusively in the realm of Ideas. Its purpose is to define concepts and to investigate their truth. By the dialectical process of "division" that which is common to several things is isolated, and a class is separated into subclasses or finally into its individual members. But as all things must go in pairs, according to the Pythagoreans, Plato's division is mated with its complementary "combination." Division and combination when fully developed would seem to be equivalent to the complete modern apparatus of cross-classification as in symbolic logic.

Animals for example, when "divided" with respect to the pair of contraries, male, female, separate into two mutually exclusive classes, each of which may be further "divided" by dichotomizing with respect to other pairs of contraries, and so on. After only 30 divisions, 153, 485, 404 subclasses are available for the pigeonholing of all animals. Each presumably would contribute its Idea to the total of Absolute Being. The initial dichotomy, male, female, for instance, might participate in the Ideas "Masculinity," "Femininity." The Pythagoreans, we saw, divided with respect to their ten pairs of contraries. The Platonic dialectic dichotomizes with respect to the categories of Reality, five of which are said to be Identity, Difference, Rest, Motion, and Being. Only dialectic, it is also said, can generate valid science.

As might be supposed Platonic realism offers certain difficulties which have not yet been fully clarified. Some were pointed out by Plato himself. How, it is asked in the *Phaedo*, can an Idea, which by hypothesis is immutable and eternal, participate at all in the transient things of this sensory world, or vice versa? Again, once we begin "realizing" the evidence of our senses in Ideas, the anticipated advantage to be gained by thus supplanting manys by ones turns out to be illusory. The Ideas necessary to accommodate the "Everythings" of Thales, Anaximander, and Pythagoras begin multiplying upon themselves at a prodigious rate, until the cardinal number of Ideas paradoxically enough exceeds the cardinal number of things. Mere quibblers in Plato's day also asked what the Ideas "Hot," "Cold" were doing when not participating in the sensations "hot," "cold." The less obviously sensual "good," "true," "beautiful" were being perceived by all the disembodied souls liberated at last from the Wheel of Birth, and these free spirits were immune to changes of temperature. Therefore for them "Hot" and "Cold" were participating in Not-Being. At this point in the ancient debate Aristotle (384-322 B.C.) injected some rather discourteous remarks. Naturally we ask how competent he was to criticize Plato's metaphysics.

The son of a physician and himself educated for the profession of medicine, Aristotle, unlike Plato, was not congenitally hostile to empirical science. For about nineteen years he was a regular attendant at Plato's lectures. From the age of twenty-one till Plato's death (349 B.C.) Aristotle was pupil, critic, and respectful admirer of the founder of the Academy. This was the period of Plato's life which he devoted to perfecting his theory of Ideas. Aristotle therefore had ample opportunity to form a first-hand, discriminating estimate of Platonic realism. Two obstacles however may have prevented him from being as objective as one philosopher judging the labors of a rival should be. Both were purely personal.

It is said that Aristotle had hoped and expected to succeed Plato as director of the Academy. When Plato died, willing the succession to Speusippus, Aristotle left Athens in a huff. But he cooled off, and on his return to Athens set up his own Lyceum in competition with the Academy. Plato knew Aristotle better than Aristotle knew himself. The industrious naturalist and crabbed logician was no fit scholar to nourish the Ideas in the Academy or anywhere else. Nor was a man who was about as insensitive as Socrates to the beauties of mathematics a promising candidate to develop the higher numerology of Absolute Being. Aristotle therefore found himself disinherited of his hopes. The disappointment and the complexion of his own intellect may have disqualified him as an entirely impartial critic of Plato's philosophy—unless he was so inhumanly scientific as to be immune to human frailty.

Aristotle's sharpest criticisms of Plato's realism were directed at its final form, that in which the Ideas are Numbers. Following Aristotle, and giving that letter forged in Theano's name all the credit to which it is not entitled, we can transfer the obscurities in Platonic realism to their origin in Pythagorean numerology. The forgery, we recall, asserted that Pythagoras said, "Things are represented by numbers," or "Things embody numbers." Aristotle declares that Plato at his least mystical meant nothing more advanced metaphysically than this unworkable antique; for "numbers" and "represented by" he substituted "Ideas" and "participate in"—a mere verbal change. But as Aristotle was unsympathetic to much of Plato's thought and is said by some Platonists to have been incapable of understanding the rest, his accusation may carry no weight. The conventional opinion is that Plato identified his Ideas with his Ideal Numbers, and that these were an invention of his extreme old age when he had lost the capacity for unmystical thought.

Aristotle himself favored the conception of (natural) numbers as "collections of units." But the appearance of irrationals had shown that irrationals (like the square root of 2) either are not generated from numbers at all, or not all numbers are "collections of units." The irrationals cannot be obtained by the addition of units, nor by finite collections of the ratios of numbers thus generated. Plato denied that the natural numbers 2, 3, . . . are the collections $1 + 1$, $1 + 1 + 1$, . . . , and asserted in effect that "they are qualitatively what they are." Certainly, he said, they are not "collections of units." A "collection" is one thing, a "number" is another. This seemed to give some meaning to his project of devising Ideal Numbers in which both natural numbers and irrational numbers might "participate."

If the earlier theory of Ideas seems obscure to unrealistic mathematicians its successor in the Ideal Numbers seems doubly so, even in Aristotle's quizzical exposition. Some of the questions Aristotle propounds sound slightly satirical, as if he were more concerned to exhibit his own superiority by showing up the late master of the Academy as a deluded mystagogue, than to gain an understanding of his ripest philosophy. Why, he asks, is a number, considered as an aggregate, one? Is this "one" the One of Platonic numerology, the One that begets All Things on the "Great-and-Small"—that mysterious shadow of the continuum which Plato left unexplained?

The question appears to be unanswerable, for Plato had assigned "mathematical objects" to a region above the many of the senses and below the One of the Ideas. Though eternal and immutable the objects of mathematics are of lower dignity than the Ideas: each Idea is the only specimen of its kind, while many mathematical objects may be alike—innumerable sensory threes, for example, but only one Idea "Three."

The argument scarcely becomes clearer as it proceeds to untangle the involutions of the Platonic Trinity: Sensory objects; Mathematical objects; Ideas. By participation in the One the "Great-and-Small" generates Ideas, and these are the same as Numbers. The Ideas are the causes of all things; all things are composed of primal elements; the elements are numbers. Numbers are generated from the "Great-and-Small," and likewise for Ideas. So the real elements of all things are both Ideas and Numbers, and Numbers are the cause of everything. But since the Numbers are Ideas they are inaccessible to the senses and should not be expected to behave as the mathematicians' numbers, which are not Ideas. The Ideal Numbers are apprehended by the reason, the mathematical numbers by the senses.

To remove the Numbers yet farther from the earthiness of useful or comprehensible arithmetic, Plato declares that they can be neither added nor multiplied. Quite pertinently Aristotle asks how one Idea can generate many Ideas, as it must if Ideas are Numbers. Plato seems to have answered him: it cannot. For if an Idea, which is a Number, were conceivable as a "collection of Ones"—as it should be if it is to generate other Numbers or Ideas—addition of Numbers would be possible. But Plato has said it is impossible. Aristotle also asks what is the precise difference between the mathematical "one" and the real "One" if the latter is a number, or Number, at all? As Plato was dead when Aristotle put this question it is still unanswered. Taking the last step possible in his universal numerology, Plato incorporated the hierarchy of Ideas under the supreme Idea "the Good." The Good thus became a Number and a Number became the Good. The limit had been reached. Number was deified.

Perhaps only a mathematical realist can fully understand the theory I have tried to outline. I am acutely conscious of the shortcomings of my attempt. It has been presented simply to give some conception of the depth and breadth of the enquiries into human knowledge provoked by the careless remark of Pythagoras that everything is number. If the little the ancient Pythagoreans knew of mathematics and its scientific applications generated a philosophy which embraced everything, from the body's awareness of hunger to the soul's knowledge of the Absolute Good, it is scarcely surprising that vastly more mathematics has enabled the modern Pythagoreans to discover the mere physical universe in their own heads.

Having confessed to one possible lack of complete understanding a skeptic might acknowledge another. A devout mathematical realist, say X.Y.Z., still bound to the Wheel of Birth, may occasionally recall some trivial fragments of a mathematical Idea, as Meno's slave remembered his prenatal knowledge that twice four is eight. He will then as a rule write out his reminiscences and send them to a mathematical periodical to be published under his own name—"By X.Y.Z." Should not the realist, if he really believes in realism, publish what he has remembered or observed under the real author's name—"By the Absolute"?

Plato's theory of Ideas was completed in the fourth century B.C. Why, the indifferent scientist may ask, should anyone in the twentieth century A.D. take it seriously? And what is to be gained by exhuming buried controversies over the validity of this or that detail of a primitive attempt to solve the universe? What possible significance can these "old, unhappy far-off things and battles long ago" have for a world which has moved forward with science? Surely the function of the history of science or of mathematics is not to preserve the obsolete from oblivion. Then why rehearse these antique debates of Plato's long-defunct Academy? Are not those philosophers who stigmatize the Platonic Absolutes as "pernicious futilities" justified in their outspokenness? Whether or not they are, no scientific mind of the reactionary twentieth century can dismiss the doctrine of Ideas as a negligible error of the past. The inveterate and implacable enemy of science is not dogmatic theology, as some scientists have supposed, but realism in the Platonic sense. It is the antithesis of science; and its popularity has increased more rapidly since 1920 than at any previous time since the sixteenth century.

Compared to the massive persistence of the refined magic and intuitive mysticism of this realism, the persecution of Galileo in the seventeenth century and the vilification of Darwin by embattled theologians in the nineteenth were passing misunderstandings of but little consequence for science. But the continued slow pressure of the sum total of all prescientific mythologies and superstitions has not diminished with time. Its thousands of years may yet overwhelm the three centuries of modern science.

CHAPTER - - - - - - - - - - - - - - 21

Pythagoras in Purgatory

HAVING witnessed the apotheosis of Pythagorean number mysticism in the Ideal Numbers of Plato, we must now follow Pythagoras himself through his purification of seventeen centuries to the Renaissance. His sufferings began in the first century B.C. with the infernal invention of Neo-Pythagoreanism by one Nigidius Figulus, a Roman illogician who started the noxious ferment of Neo-Platonism by infusing the Platonic Ideas with oriental mysticism. From that torture Pythagoras descended to the chaotic philosophical Hades of Gnosticism. But he was not to remain there forever. Aided by able and sympathetic Fathers of the struggling young Christian Church, he overcame the Gnostics and began slowly to rise through the mephitic vapors of decomposing philosophies. Passing from the Dark to the Middle Ages he proceeded on his arduous ascent to seventeenth-century science, enduring mediaeval sacred and profane numerology as he rose. In this period of his sojourn in the underworld of science and sanity his agonies were extreme. At last, in the fifteenth century, he encountered Plato, who also was on his way up after having been hurled into Tartarus by Aristotle. Together they decided to make a dash for freedom.

They escaped just in time to witness the birth of modern science. Seeing nothing to do for the moment the two sages, now one in friendship, parted, agreeing to meet in 1920. Shattered in spirit and mentally exhausted after all they had gone through both longed for rest. Pythagoras recuperated in mathematics, Plato in metaphysics. The fateful year 1920 found them refreshed and eager to continue the collaboration they had begun in purgatory.

It would be harassing to detail the extravagances and excesses of the pure reason in its unrestrained revel through Neo-Pythagoreanism, Gnosticism, Neo-Platonism, and the theological numerology of the Middle Ages. Nor is it necessary for us to do so. The ancient Pythagoreanism with which we are already acquainted lived again in weird reincarnations all through these masterworks of the unaided human intellect. It will be sufficient to indicate the general characteristics of each main period and cite a few names familiar to nearly everyone as those of some of the greatest pure reasoners our race has produced.

Through much of this triumphal progress of the pure reason "the queen of the sciences" was astrology. In the Middle Ages astrology shared her throne with theology. Not till the nineteenth century were both astrology and theology dethroned by Gauss to make room for mathematics. These three rulers faithfully represented the best that pure reason had to offer in their respective reigns. Their rise to domination over the minds of acute reasoners endlessly explaining the universe to a docile and patiently credulous humanity, and their subsequent decline as substitutes for those despised drudges, observation and experiment, consumed about four-fifths of the time span from Nigidius Figulus to Albert Einstein. In contrast with this protracted despotism of the pure reason, modern science has governed the thoughts and guided the actions of a numerically negligible company for approximately three-twentieths of the same span—about one-fifth as long as the other. Four to one would seem to be generous enough odds in favor of the unaided reason. Yet even this heavily weighted advantage failed to stimulate any material achievement comparable to what modern science shows in a week.

But possibly, it may be asked, the immaterial gains from the mediaeval conception of the good life were so overwhelming that mere conveniences in living and in understanding the world in which we live are comparatively of no account? The Middle Ages, we are constantly reminded, have been rehabilitated, and the thirteenth century is now recognized as the golden age of the Christian era. And from several trends since 1920 it may not be too optimistic to expect a similar restoration of the Dark Ages before the twentieth century is out. When sympathetically—or numerologically—viewed that troubled period also has its attractions for the nostalgic soul averse to science. Perhaps a survey of the achievements of the pure reason from 100 B.C. to 1600 A.D., more extended than is possible here, may enable those who hesitate whether to return to the past or to remain in the present to make up their minds. Here we can only follow Pythagoras, sampling his milder torments as we go.

The Neo-Pythagoreans flourished from the first century B.C. through the second century A.D. Though they were gradually superseded by less inconsistent reasoners, their own peculiar fantasies survived for many a century in the numer ology of their successors. Attempting the impossible they sought to fuse whatever captivated their undisciplined imaginations in the philosophies of Plato, Aristotle, the Stoics, the ancient Pythagoreans, and all oriental mysticisms into a superphilosophy of everything in heaven, earth, and all the hells revealed up to their time. They were, they claimed, the legitimate inheritors of the Pythagorean mysteries. When oral warrant for their pretensions failed to convince doubters they resorted to forgery, producing letters and treatises signed with any name, including Theano's and the master's, that might impress the credulous. In their way of life they endeavored—not too successfully—to observe the rigid discipline of the legendary Brotherhood.

Intellectually they vacillated between uninformed enthusiasm and conscious quackery. The cement holding together their jumble of inconsistencies in a crazy parody of consistency was the original Pythagorean numerology eked out with scraps of its perfection in Platonic realism. The same Limited and Unlimited, the same male One and the same female Two that had distracted the ancient Pythagoreans once more divided and ruled the universe between them. But now, as was inevitable after Plato had lived, these hoary Numbers were less anthropomorphic, more metaphysical, than when Pythagoras whipped them through their tricks.

To the moonily tolerant Neo-Pythagoreans nothing that could be said about numbers was absurd, and they said nearly everything a demented numerologist might shout in his delirium. Nor was any metaphysical impossibility beyond proof by their mystical magic of numbers. Ominous rumblings of the debacle of Pythagorean numerology that was to come in the sacred number mysticism of the mediaeval theologians were clearly audible in the prolonged controversy, rashly revived by the Neo-Pythagoreans, between the impeccable Monad and the incorrigible Dyad. As in the ancient numerology the male Monad, the One, was all-good, all-wise, all-knowing, eternal, invariable; while the female Dyad, the Two, was the source of all evil, foolish, ignorant, transitory, unstable. The Monad symbolized the deity, the spirit, perfect form; the Dyad was the mark of matter, the senses, chaos. Some centuries later the Two surreptitiously changed her sex and became the devil. For that unpardonable lapse of taste the unfortunate Dyad was damned beyond all hope of redemption by the exasperated Monad.

Pythagoras did not suffer alone in these excesses of his self-chosen disciples. Plato also was taken apart and put together again in the higher transcendental nonsense, and Aristotle was smothered in his own logic. The supreme contradiction of the revised numerology made the Platonic Ideas objective realities as massive as the pyramids of Egypt, and at the same time thoughts in the mind of the deity as insubstantial as the dreams of a butterfly. This promising absurdity was bequeathed to the more intellectual Gnostics with the Neo-Pythagoreans' blessing.

The Neo-Pythagoreans seem on the whole to have been rather inoffensive folk, doing their muddled best to fake a numerological synthesis out of an inherited welter of mutually inconsistent religions and contradictory superstitions. The aim of the Gnostics appears to have been roughly the same. But if we may judge by what the Christian Fathers said of them, the erudite Gnostics were conspicuous primarily for their conceit. Though their authority had begun to wane by the middle of the third century, they were the highbrows of the first to the fifth centuries of the Christian era. They were also —though they were unaware of the fact—the last frantic spasm of an exhausted philosophy to live. Alexandria, the city that had sheltered the climactic school of Greek mathematics, became their asylum.

The title that satisfied their predecessors in the pursuit of wisdom was not honorific enough for these decadents. Instead of styling themselves philosophers they adopted the nakedly pretentious "gnostics"—those who know—to describe their own peculiar merits. Their supposed absolute knowledge changed complexion like a chameleon with the color of its surroundings.

No clearly defined system can be attributed to these aimless eclectics to whom Babylonian astrology was as acceptable as Platonic theology. Like the Roman Empire in whose incipient twilight they basked, the Gnostics welcomed all gods, all religions, all superstitions, all "sciences" in so far as they were unscientific, and all theogonies into their panmixia of knowledge and nonsense.

The one feature that gave the Gnostics' miscellany of contradictory myths and superstitions a semblance of coherence was the ancient Pythagorean numerology as debased by the Neo-Pythagoreans, and that itself was incoherent. It became more so when gematria, the Hebrew variant of numerology, joined the melee.

Jewish number mysticism has always had an unfair advantage over all others since Plato, in that letters of the Hebrew alphabet are used to write numbers. Consequently any passage in the Talmud has at least two meanings according as it is interpreted in words or in numbers. The numbers when properly manipulated generate others, and the results are interpreted in words. The magical possibilities are plainly infinite, and it is a conservative guess that vastly more esoteric truth has been discovered numerologically in the sacred writings of the Jews than in any similar body of literature.

Incubated in Gnosticism, and hatching out fully in the mediaeval numerology of both Jewish and Christian theologians, gematria evolved into the most flexible of all number magics. In the perfected form numbers chosen at will were assigned to the letters of any alphabet. By puerile arithmetic and elementary cheating almost any word was thus forced to yield almost any desired meaning, and it became a matter of simple routine to curse an enemy by blessing him, or vice versa. Mere absurdity or self-contradiction was nothing against a particular deduction. If their numbers identified the reigning Pope and Satan, or Christ and the Antichrist, the astounding revelation was but another proof that the true and the incomprehensible are one.

What can have induced the Gnostics and their successors to follow the vagaries of arithmetic so devotedly? Nobody knows. Even less restrainedly than the ancient Pythagoreans these deluded pundits embroidered every trivial relation between numbers with fantastic mysteries. Turning their backs on common sense they pursued a meaningless mysticism of numbers through one absurdity to another, humbly believing in all and astonished by none. A modern mind encountering some of these horrors without previous knowledge might imagine them to be the play of lunatics. Not at all: the wildest excesses of nonsense were not the sport of idle pranksters or witless jesters. They were the sober work of conscientious theologians and the closest reasoners of their respective ages.

We who defer to science in matters of ascertainable fact find it difficult to believe that less than five centuries ago human beings like ourselves lived and died by the rules of an insane arithmetic. Even the dead were not left in peace. Their time in purgatory could be lengthened or shortened by the appropriate reading of the numbers on their tombstones. And all this magic of numbers was the very heart of knowledge and the essence of wisdom. Pythagoras knew much, but he knew nothing like this.

Compared to Gnosticism and its outgrowths in mediaeval "science" the ancient numerology was the soul of modesty. Infinitely more than the Pythagorean "everything" was number for the Gnostics and their erudite kind all down the Middle Ages and well into the Renaissance. Impossibilities existing neither in the material world nor in the memory of the deity were numbered and reckoned with the rest. Learned Europe became the madhouse of arithmetic.

When the amiable knowers of all things knowable and unknowable saw that Christianity was acquiring an intellectual following that would sooner or later have to be reckoned with, they cheerfully welcomed the young religion into their menagerie of wild cults and half-tamed creeds. But the rugged Christian Fathers would have none of this aggressive hospitality. They denounced the Gnostics as a mob of degenerated Greek philosophers taking the names of Pythagoras, Plato, and Aristotle in vain in their preposterous botch of Pythagorean numerology, Platonic realism, and Aristotelian categories—to say nothing of the Egyptian trinity of Horus, Isis, and Osiris, or of the Persian duality of body and soul, or of the astrology of all peoples and all times. Let these pretenders to divine knowledge swear by Pythagoras as the god who had willed them the sacred tetractys, till the One became Many and the Many became infinitely More: the staunch Fathers of an unpretentious young religion refused to be impressed by anything—with one exception—the learned Gnostics might promise or offer. That exception, unfortunately for the sanity of ten tormented centuries, was numerology.

As Plato and Pythagoras share about equally in the honor of restoring the physical sciences to numerology in the twentieth century, we must note briefly what befell Plato while the later Gnostics were torturing Pythagoras. In a word it was Neo-Platonism. The natural offspring of Neo-Pythagoreanism and Gnosticism, this unstable compound of crude numerology and mystical metaphysics originated with Plotinus (205-270 A.D.). An austerely genial mystagogue, Plotinus came out of Egypt to settle in Rome and undertake a universal salvage of pagan philosophies four centuries too late. (It is sometimes asserted that Plotinus' teacher, Ammonius Saccas, was the real founder of the "school" of Neo-Platonism. The difference in time is inconsiderable; and whatever glory there may be in having fathered such a hodge-podge is scarcely worth a squabble, even between professional scholars. The appalling fact that Neo-Platonism actually happened to the human mind is enough for any modern scientific observer to note and remember, lest the like overtake it again.)

Neo-Platonism has been called the third and last period of Greek philosophy. Considered simply as philosophy, Plato would scarcely have recognized it. Aristotle might have tossed it a contemptuous greeting—"Just what might have been expected to come out of old Plato's Numbers"—as he passed by it on the way to his prolonged triumph in the Middle Ages.

Confusing Judaism, Hellenism, and oriental sciences and

religions in one sublimely inconsistent whole, the Neo-Platonists struggled first and last to explain the ancient dualism of appearance and reality. Plato, of course, had said much on the same subject. His self-styled successors said a great deal more. They talked themselves out of their senses and into a total mysticism in which subject and object became one, and knowledge was possible only by coalescence with the deity. Greek philosophy as a guide to sane living expired. But it was not to be accorded honorable burial. From an intricate theology of polytheism, Neo-Platonism proceeded with its own parody of the Platonic dialectic to a systematically confused synthesis of all classical philosophy. Logic went mad.

Of all the charlatans, magicians, self-intoxicated mystics, and ecstatic logicians who made Neo-Platonism what it became after Plotinus, we need mention only Proclus (411-485) of Constantinople, Alexandria, and Athens, precursor and mystical inspirer of the more philosophical Christian numerologists of the Middle Ages. Undoubtedly Proclus was a great man by almost any criterion uncontaminated by science. His life was that of the aggressively pious but otherwise blameless enthusiast who has beheld a blinding vision of the one true philosophy, and who ever thereafter insists on putting out the eyes of those who can still see. His differences with the powerful Christian authorities of Athens caused both them and himself considerable discomfort. Practical ethics made no appeal to Proclus. He demanded mysteries, and he discovered them in abundance in the decaying remains of Neo-Pythagoreanism and a belated resurrection of the Orphism which preceded Anaximander. Soon he found it as easy as thinking to summon beneficent spirits to aid him in his divinely imposed task. His destined mission, at which he labored prodigiously, was the seduction of Christian converts to his own subtly exciting numerology of nature and the human soul. Claiming magical powers for his meaningless formulas he boldly advertised his pretended control over spirits and the material world, and hinted that others might exercise similar or even greater powers. Provided only that they had sufficient faith mere mortals could compel the very gods to do all their drudgery for them. A slip, of course, might raise the Dyad, but then the stakes were high and the risk not too great. This Arabian-Nights substitute for what the Christian teachers offered proved irresistible to the weaker proselytes, and Proclus found himself unpopular with his influential rivals. They threw him out. On being forgiven after a short exile Proclus returned to Athens, more obstinately pious than before in his own perverse way, but also more discreet. He talked less and wrote more.

In his higher arithmetic of the soul Proclus inaugurated the scientific method, so-called, of the Middle Ages. For a thousand years numerology reinforced by an intricate dialectic—disrespectfully termed logic chopping by unsympathetic modern scientists—fought savagely to usurp the functions of observation and experiment. Beyond this historic triumph Proclus has little further interest for us. A thousand years' survival in the errors of his fellows should be immortality enough for almost any man; and it does not matter much any more to anyone exactly how Proclus derived his unique Supernal Number from his three Absolute Ones by triadic involution, evolution, and emanation from his Original Essence.

As paganism gradually gave place to Christianity, the Pythagorean number mysticism changed its objective but not its fundamental technique. Accepting number as the supreme authority in what passed for science, ecclesiastical scholars elaborated their own perversions of the ancient numerology as aids to understanding the holy scriptures and also, it must be said in the interests of theological honesty, as proofs that the scriptures are true revelations of the divine word. Thus in mediaeval Christianity as in ancient Pythagoreanism and some aspects of Platonism, number was more powerful than the deity. But not always; numbers frequently masqueraded as divine creations. Never were they man-made—which need not surprise us when we remember what twentieth-century mathematical realists believe about the nature of mathematics.

Before passing to a few particulars we remark once for all that however ridiculous the numerology of some of the great men cited may seem to a modern mind, those men nevertheless were great. Competent judges estimate at least three of them—Augustine, Albertus, Aquinas—as the intellectual peers of the greatest men of any age. Their numerology was only one phase of their fervent activity; and if it seems strange in a scientific age that these giants of the past should have taken number mysticism with the devastating seriousness they did, it may seem even stranger to our successors a few centuries hence—should Pythagoras decide to prolong his stay with the mathematical physicists and astrophysicists—that we accepted empirical science without a doubt or a smile. The most disturbing thing about the future is that nobody can foresee who will be next undone.

Numerology as an orthodox method of research in mediaeval theology stems from St. Augustine (353-430), "a man of towering intellect" according to believers and infidels alike. Born a pagan, Augustine retained some of his joy in living even after he had become the outstanding champion of his acquired religion. "O Lord," he petitioned, "make me chaste, but not quite yet." After an enthusiastic study of Plato and a loving reading of holy writ, Augustine applied his powers to getting numerology accepted as the basic science supporting Christian theology. Number for him was the very essence of truth and reason. So if the scores of ones, twos, threes, fours, sevens, and all the tens, forties, and even richer numbers in which the scriptures abound, were correctly interpreted, the reasonableness of the accompanying theology would be established beyond all cavil. Accordingly Augustine made an exhaustive numerological analysis of the whole Bible.

The fault—if it was one—was not Augustine's that many of the meanings he thought he detected in even the most casual mention of numbers were as forced and as fantastic as any absurdity of the Neo-Pythagoreans. It was his implement that was to blame. When he rose to the more philosophical levels of number mysticism his findings agreed substantially with those of Plato and the modern Pythagoreans. "It is clear to the dullest intelligence," he declared, "that the science of number was not created by man but was discovered by investigation." From that obvious truth and his numerology of the scriptures he concluded that number is the unshatterable foundation of the Absolute, and that the deity is the Great Numerologist, who knows all numbers because his understanding is infinite. Conversely, the deity is omniscient because he knows all numbers. Number is therefore necessary and sufficient for the existence of the deity.

Not all of the more extravagant deductions were Augustine's

own. Many were incorporated into his comprehensive analysis from the labors of earlier Christian thinkers, most of whom had yielded to the charms of Gnosticism and Neo-Platonism. The esoteric doctrine of the sacred three, for example, was already highly developed when Augustine took it over, amplified it, and passed it on enriched by his own contributions to the theologians of the Middle Ages. The central difficulty here was to show that Pythagorean numerology sanctions the equality $3 = 1$. By the time Augustine attacked the problem of the three this difficulty had been overcome. The Council of Constantinople (381 A.D.) had officially recognized the transcendental arithmetic of the holy trinity as the foundation of Christian theology. But the Council might have hesitated long to endorse all of the conclusions their successors drew from that prolific postulate that three and one are the same.

It may be left to the imagination what was done with the three gifts of the magi, St. Peter's thrice-repeated denial of his Lord, the three days between the crucifixion and the resurrection, and the three appearances of the risen Lord to his disciples. Some of the deeper theorems might strike a modern theologian as slightly blasphemous. But, for their time, they were not. They were the sincere efforts of reasoning men to convince themselves that the scriptures are divinely true, that nature violated herself, and that miracles did happen. Nor is it remarkable that these believing men sought to support revelation by appealing to the only "science" they knew—numerology, when the like is common today with the latest instead of one of the oldest of the sciences as the court of appeal. It seems never to have occurred to Augustine and his able followers to enquire whether it was their religion or their science that they really believed in. The human mind is more wonderful than nature.

What the theologians left intact of elementary arithmetic was maltreated and debauched by the alleged mathematicians. Compared to those who preceded or followed them, the mathematicians of early and mediaeval Christianity were rather sorry specimens. Practically all of them were devout believers in number magic. Borrowing their method from the Neo-Pythagorean Nicomachus of the first century A.D., whose ignoble classic on arithmetic retailed much Pythagorean numerology in an attractive package, the majority of scholars paid more attention to the supposed mysteries of numbers than to the practical side of arithmetic. All appear to have been convinced that number is the key to all sciences and all philosophies, and not one of them doubted its celestial origin. Only a few names, well known in other connections, need be recalled as typical of the best numerological thought of their times.

Boethius (480-524 A.D.) was the last considerable Roman scholar who understood Greek. His *De Consolatione Philosophiae*, composed in prison, is still cherished by those few happy mortals whom philosophy can console. The high ethical tone of this famous tract has caused scholiasts to suspect that its author was not a pagan, and therefore not Boethius. But the authenticity of both the authorship and the paganism is well established.

In his numerology Boethius adhered to the pure faith of the master himself. "All things," he declared, "do appear to be formed of numbers." But he was not by any means all mystical arithmetician. His elementary manuals of arithmetic, astronomy, geometry, and music—the four Pythagorean sciences—depressed European education clear through the Middle Ages. And his translations of as much of Aristotle as he knew were long the only direct connection between classical Greek philosophy and mediaeval theology. Undoubtedly these translations were largely responsible for the protracted tyranny of the Aristotelian philosophy over the minds of clerics and secular scholars alike. If Boethius had also succeeded in transmitting some of Plato's dialogues the story of European culture might have been quite different from what we know.

After serving the Gothic King Theodoric faithfully and competently as court minister and consul, Boethius was executed in a needlessly brutal manner. Officially the charge was treason. Unofficially Boethius was put to death because he was incorruptible.

Dying about a century later than Boethius, the polymath St. Isidore (570-636), Bishop of Seville, continued the propagation of the Pythagorean gospel in an extremely erudite encyclopaedia of the numbers occurring in the holy scriptures. The Bishop's ingenious lucubrations did much to fix the style of numerological exegesis of the Bible in the works of his innumerable successors down to Dante in the thirteenth century. Following St. Augustine, Isidore ably expounded the cardinal doctrine of the ancient Pythagoreans that all things are contained in the Decad. Since the Decad is generated by the Monad, it followed precisely as it had for the Neo-Pythagoreans that everything is—or should be—eternally One with the deity.

Where meaning is fluid, controversy flows like water. In retrospect mediaeval numerology appears as a continuous and futile war of words between innumerable antagonists all saying the same thing and none meaning what any of the others meant. Solemnly or bitterly debating whether this or that ridiculous interpretation of some sacred ten or thrice-sacred nine was in accordance with the true Pythagorean principles, the heated doctors of divinity overlooked the one question of any significance. None of them ever stopped arguing the merits of his own favored mysteries long enough to ask himself whether numerology itself might not be meaningless. Possibly it was not, in the Middle Ages. Meaning, like beauty, may have been only a matter of personal taste to the busy unifiers coordinating the fortuitous numerical coincidences of nature, philosophy, and holy writ in a universal and incomprehensible One. When more of Aristotle's works than the Boethian translations became available to European scholars, the great naturalist-logician's disguised Pythagorean numbers—four causes, four elements, ten categories—imposed their richness on the already surfeited confusion.

Aristotle's authority being second only to that of holy writ itself, his logic and his account of the universe united with Christian numerology and theology to rule the reason in a dual despotism no orthodox scholar, theologian, or scientist dared defy. Even intellects of the stature of an Albertus Magnus (1193-1284) and his superhumanly logical pupil, St. Thomas Aquinas (1226-1274), submitted. Their authoritative example determined the main current of European thought concerning the physical universe and man's relation to it for three blighted centuries.

Experiment was not absolutely neglected during even the least scientific decades of the tyranny of the "pure reason." But it was haphazard and, with but few exceptions, negligible in both quantity and quality. Against the torrent of words gushing from hundreds of eloquent logicians and loquacious numerologists all solving the universe entirely in their heads, European science had all it could do to remain stationary and avoid being sluiced back to the Stone Age. Even Roger Bacon, towering above the flood like a Gibraltar, though not swept away, was finally submerged and forgotten till human beings, quite literally, came to their senses. Using their hands and their eyes, they discovered that all that deluge of thrice-distilled reasoning was a bad dream that had suddenly vanished in the dawn of modern science.

Hints of what was to come are plain enough now in the life and works of Bacon (1214-1294), contrasted with those of his more famous competitor for remembrance, Dante Alighieri (1265-1321). Contemporaries for almost thirty years, Bacon and Dante were as discordant a pair as history ever turned up in the same century—and that the golden summer of the Middle Ages. Both knew trouble in many shapes, and each imagined his own "revelation" provided the one true approach to life. The poet was as ancient as Pythagoras, the scientist as modern as Galileo. The natural philosophy of one was to linger on for about two hundred years before it was buried forever; that of the other slept for nearly three centuries before it came fully to life. Dante was the Middle Ages incarnate; Bacon was the unembodied spirit of the age of modern science. Neither was recognized by any of his contemporaries for what he was, still less for what he was to become. Dante achieved quick and lasting reputation. Bacon has had to content himself with the empty honor of having been a "might have been"—had he been born three centuries later than he was. Few but specialized scholars of Italian literature today know anything of Dante's mystical substitute for science, and only a rare illiterate believes a word of it; while millions are alive, if nothing more, only because the experimental-mathematical science which Bacon tried too soon to teach the world, at last was taught and accepted.

Numerology had its poet in Dante. Embroiled in cut-throat Florentine politics for much of his early life, and a hunted exile in his prime, Dante yet found leisure not only to compose one of the world's great poems, but also to make himself complete master of the philosophy, the theology, the astronomy, and the physical science of his age. In number mysticism he was the accomplished artist and the learned expert without a peer. If anything worth preserving for its own perfection could be made of numerology, Dante was the man to make it. And he did. Himself saturated with both ancient and mediaeval number mysticism, he could hardly have avoided expressing his esoteric philosophy of heaven and hell in the symbolism of numbers, even if he had wished to conceal his art. But he chose his medium deliberately, knowing that others as learned as himself would find the deeper meanings hidden in the numbers of his *Divine Comedy*. As a matter of fact it required no great learning to follow the close interplay of theology, human and divine love, and mediaeval cosmology in the "angelic nine" recurrently associated with "the mystery" Beatrice. Anyone with any schooling in Dante's time was as familiar with the sacred implications of one, three, and three times three as children of a generation ago were with the multiplication table. Scholars might discover a feast for the pure reason in the ancient numerology fused with the new, but the unlettered would find food—perhaps not unmixed with poison—for their souls.

The rich and the poor in knowledge for whom Dante wrote died with the Middle Ages. The numerical symbolism in which he embedded his narrative with the consummate skill of the master lost its significance long since, and only the poetry remains. Yet, had not Dante consciously worked by a pattern that now has no meaning, his poetry might also be no longer readable.

In stark contrast with Dante's achieved success, Roger Bacon's frustrated life mirrors the thirteenth-century conflict between "two worlds, one dead, the other powerless to be born." This, it may be emphasized at the outset, is only one of two current estimates of Bacon's life. It is that, apparently, of a majority of modern scientists acquainted with the relevant parts of Bacon's writings and with those details of his life which are undisputed.

The contrary estimate presents Bacon as a greatly over-rated compiler and encyclopaedist, a vain egotist, and a perpetual grumbler who could not possibly have suffered any inconvenience on account of his original and advanced views, for the adequate reason that his opinions on scientific matters and mathematics were neither original nor advanced. According to this deflation, Bacon merely parroted his more enlightened contemporaries and some of his mediaeval predecessors.

It is undeniable that Bacon himself made no contribution to mathematics, and that some of his proposals for scientific experiments are ridiculous. His conception of the scientific method and the part which mathematics might play in science will appear as we proceed. As for the admitted absurdity of some of his proposed experiments, it is interesting to compare Bacon's proposals with those of the scientific gentlemen who frequented the meetings of the Royal Society of London for the Advancement of Science shortly after its incorporation in 1662—when Bacon had been dead for 368 years. Between some of those sober offerings to the scientists of the seventeenth century and Bacon's projected experiments there is not always a great deal of difference.

There seems to be an excess of zeal on both sides of the dispute. The champions of the Middle Ages insist that mediaeval science, especially its experimental aspect, has been grossly misrepresented. The proponents of modern science retort that whatever virtues the Middle Ages may have exhibited, in science or in anything else, the world had enough of the mediaeval mind in the past, and wants no more of it either now or in the future. And squarely in the middle of the controversy stands Roger Bacon, indifferent to the prejudices of admirers and disparagers alike. Though he was neither Galileo nor Newton, he knows—if his spirit longer is aware of anything—that Galileo and Newton would have welcomed him to their company.

To say that Bacon was in some respects no more advanced than his contemporaries is to utter a platitude. Why, in particular, should he be censured for not rejecting all the absurdities of his age when some of them persisted for centuries after his death? A far better mathematician than Bacon could ever have hoped to be was a more devout number mystic than Bacon ever was, and that three centuries after Bacon was dead. Kepler surpassed Bacon both as a numerologist and as an astrologer. Although the vitality had already gone out of the Pythagorean numerology and all that issued from it when Bacon was born, it still rolled blindly on in time by the sheer inertia of seventeen centuries of tradition, crushing any independent mind that rose up to halt its moribund authority. Only when Pythagoreanism was temporarily ignored, but not stopped, did modern science begin to live.

Naturally, being a man of his own age, Bacon paid the ancient tyranny the tribute of his sincere respect: "Mathematics is the gate and key of the sciences, which the saints discovered at the beginning of the world . . . and which has always been used by all the saints and sages more than all the sciences. Neglect of mathematics works injury to all knowledge, since he who is ignorant of it cannot know the other sciences or the things of this world. And what is worse, men who are thus ignorant are unable to perceive their own ignorance and so do not seek a remedy."

Wrenched from its context (in the *Opus Majus*) and severed from its epoch, this famous and much-quoted testimonial to the merits of mathematics might pass for little more than a harmlessly inflated statement of historical fact. Except for its inclusion of the saints—Augustine and his successors—Bacon's tribute might be endorsed by almost any follower of Galileo and Newton. But Bacon's words did not mean for him what they mean for us, as is evident from his perfectly sober comments on the mystical numbers of astrology. Although mathematics might be "the gate and key of the sciences," it was astrology whom the mathematician found reigning as queen after he had unlocked the gate and entered the kingdom of science. The mediaeval fraction of Bacon's mind bowed to astrology, the rest was free.

After some years at the University of Paris, where he studied the sciences and languages, including Arabic, Bacon returned as a lecturer to Oxford, where he had been a student. There he struggled unavailingly to replace logic by mathematics in the university studies. "Divine mathematics," he declared, perhaps unconsciously plagiarizing Plato, alone could form a sound basis for education, for only mathematics "can purge the intellect and fit the student for the acquisition of all knowledge." If Bacon's eulogy was intended as an indirect slur on the interminable quibbling of the Aristotelians in whose midst he attempted to practice science, it was well aimed. Defying the bigoted logicians among his colleagues he boldly preached the heretical doctrine that experiment is the one reliable foundation for the natural sciences. Moreover, whether or not he practiced what he advocated, he unequivocally stated the modern scientific method of proceeding from the mathematical formulation of empirically discovered principles to deductions from them, and comparison of the results with observation or further experiment. Nor did he disdain utility as a motive in science. The few who heard what he said learned nothing, possibly—though this seems unlikely to Bacon's scientific admirers—because it all was already familiar to them.

Bacon's premature science was born of his knowledge of what the Moslems had been doing while the Europeans were losing themselves in "that haze of words in which we all drowse" till harsh fact rudely wakens us. While the Christian followers of "all the saints and sages" were arguing about the sacred mysteries of numbers, the followers of the infidel prophet Mahomet were cultivating empirical science and mathematics. Bacon could not make up his mind whether to follow the saints and sages or the prophet. He divided his mind and followed both.

Having spent a large fortune in buying books, apparatus, and Arabic manuscripts, Bacon in his early forties found himself bankrupt and almost friendless. He became a Franciscan friar. To break the monotony of the uncongenial life he continued his science. Chemistry, optics, and the search for the philosopher's stone in the mazes of alchemy, helped to relieve the tedium. Naturally enough his brother Franciscans accused him of commercing with the devil. His noisy and malodorous experiments with gunpowder undoubtedly were partly responsible for the charge.

In order that he might be properly watched, Bacon was ordered to Paris. There, Guy de Foulkes, whom Bacon had known in England, encouraged him to continue a comprehensive exposition of his scientific ideas.

When de Foulkes became Pope (Clement IV), Bacon managed somehow to scrape together sufficient funds to buy writing materials and to borrow books. In fifteen months he completed his *Opus majus*. Hopefully he sent it to Clement. As the Pope was mortally ill at the time (about 1267), and died shortly after (1268), he probably did not read the great work he had requested. But he made it possible for Bacon to return to England. This grace however did Bacon but little good. After the death of Clement, the experimental philosopher seems to have lacked both encouragement and friends. His works had already been placed on the forbidden list. Quite correctly the culprit was accused of propagating "suspected novelties." His "magic" was in fact too new by about three hundred years.

Although his chemistry was mostly alchemy, and his "divine mathematics" not wholly clean of numerology, Bacon in his larger fraction was a modern scientist. His work in optics—the laws of reflection and less precisely those of refraction, an attempted explanation of the rainbow, and experiments with magnifying glasses—alone lifted him into a world far above the mire of words in which a majority of his European contemporaries still groveled. And this holds whether or not Bacon had learned most of his optics from the Moslems. Others had the same opportunity. We do not have to believe in the historical veracity of Kipling's story, which reconstructs Bacon's emotions on seeing the protozoa in a drop of water, to credit his scientific inclination to use his senses and whatever apparatus he possessed to supplement his reason. If others did likewise, the more glory to them, but not the less to Bacon. It is a curious fact that it is impossible to raise a hundred by dragging down one.

The last twelve years of Bacon's life have been a source of controversy. One side claims that he spent them in prison, or at least in confinement, being released only a few months before his death at the age of eighty. The other denies that he was ever subjected to restraint, but fails to account for the dubious twelve years.

Whatever may be the fact, the unruly friar's tampering with the rainbow alone would have been enough to condemn any man in the thirteenth century to something even less pleasant than imprisonment for life. The rainbow was the "token of the covenant" that there should never again be such a deluge as the flood that sent Noah into the Ark. The good theologians of Bacon's time all died before they learned the elementary lesson their successors mastered so painfully in the nineteenth century: the deity, in addition to being a mathematician, is also a scientist.

About a century and a half after Bacon's death the fall of Constantinople (1453) to the Turks marked the beginning of a new era of European culture, in which a suitably refined number mysticism shared with a civilized literature, science, and art. Forgotten masterpieces of Greek learning found their way to Italy with the scholarly refugees expelled by the Turks. The end of Aristotle's long tyranny was in sight; Plato began to live again. And with the revival of Platonism, numerology for all but bigoted ecclesiastics and the mass of the people gradually became more metaphysical. Only a year after the discovery of America (1492) the common man received the first authoritative exposition of Christianized Pythagoreanism in *The Kalendar and Compost of Shepherds*. This widely popular jumble of astrology, theology, and numerology raked together as much of the authorized "science" of the time as those responsible for the eternal salvation of mankind felt they could safely sanction. A little knowledge might be the dangerous thing Pope said it is, but the authorities of 1493 knew that nothing was so dangerous to themselves as more than no knowledge at all. The number mysticism that might have died of old age when Columbus discovered a new world was kept alive by artificial inspiration.

But with the precursors of modern science it was different. Under the stimulus of Greek science and mathematics suddenly injected into the body of a pseudo-science that had lived a thousand years too long, a retarded corruption was at last released. It efficiently did its work. By the middle of the sixteenth century no reputable scholar was taking the absurdities of mediaeval Pythagoreanism and the scholastic refinements of Aristotelianism seriously. Number mysticism in the writings of the learned had returned to its Platonic perfection. Struggle as they might to restrain thinking men from rejecting the mysteries in the magical arithmetic of the Middle Ages, "all the saints and sages" were impotent to force the *Kalendar* or anything like it on the mathematicians and scientists of the Renaissance. But neither were those recalcitrant thinkers able to loose themselves from the strangling grip of the past and walk forward free men. They still saw only as much of the world as the ancients permitted them to see. Their eloquent eulogies of number as the key to the universe might have been declaimed by Plato himself in the fourth century B.C. Two specimens, both famous, are typical of the more temperate. They have been chosen rather than others perhaps less immoderate because their authors were men of interest as human beings.

The first is from that eccentric necromancer John Dee (1527-1608) of London. As an enterprising undergraduate at the University of Cambridge, John earned himself the title of magician by his anticipation of Hollywood stagecraft. The occasion was a play by Aristophanes, which the ingenious John as director enlivened with the tricks of lighting, fire and brimstone, trapdoor entrances and exits through the roof, that we now associate with a supercolossal feature by any of the major studios. John knew that he was hoaxing; his audience did not. The damning stigma of "magician" stuck to him till four years before his death. At the age of seventy-seven he was solemnly exonerated once for all by the infamous Star Chamber of the suspicion of having practiced black magic. He had been accused of attempting to abate Queen Mary's bloody life by sorcery. As Dee made no secret of his researches in alchemy, astrology, theurgy, occultism, and Rosicrucianism, it is not remarkable that his orthodox contemporaries of both State and Church looked behind him for his master Satan. Actually he was a harmless pedant, or at worst a government spy, who took himself with greater seriousness than his contributions to the propagation of learning and misinformation warranted. Describing his daily round, he announced his resolve "only to sleep four hours every night, to allow to meate and drink (and some refreshing after) two hours every day, and the other eighteen hours (except the tyme of going to and being at Divine Service) being spent in my studies and learning." A man who imposes a schedule like that on himself usually does much and accomplishes little. So it was with Dee. He and many others of his scholarly tribe rendered astronomy good service by their courageous championing of the Copernican theory of the solar system against the desperate opposition of the old guard of Aristotelian theologians, but themselves took no significant step forward. Like Dee, they would have made admirable members of the Pythagorean Brotherhood. But instead of ascribing all their opinions to the master, they gave the glory to "the ancients"—meaning all the Greeks from Pythagoras to Nicomachus. Science had to outgrow this undiscriminating reverence for the past before it could find its own way to the heart of nature.

Dee paeaned his Platonic eulogy of the ancient Pythagoreanism in his preface to the first English translation (1570) of Euclid's *Elements* (spelling modernized): "All things (which from the very first original being of things, have been framed and made) do appear to be formed by reason of numbers. For this was the principal example or pattern in the mind of Creator." Incidentally, Dee confused the geometer Euclid of Alexandria with the philosopher Euclid of Megara under whom Plato studied argumentation. This may account for the pure Platonism of his dithyramb on mathematics as a whole. "A marvelous neutrality have these things *Mathematical*, and also a strange participation between things supernatural, immortal, intellectual, simple and indivisible, and things natural, mortal, sensible [sensory], compounded and divisible . . . Only a perfect demonstration, of truths certain, necessary, and invincible, universally and necessarily concluded, is allowed as sufficient for an argument exactly and purely mathematical."

It is not necessary for us to understand Dee's numerology. On the whole he was a sound Pythagorean of the more metaphysical type. It is his panegyric on Euclid's *Elements* as the embodiment of all the logical and mathematical perfections he enumerates which is of interest. How far he was off will appear when we come to Saccheri in the year 1733. The Moslem mathematicians of the Middle Ages had as much reverence for Euclid as the British and Continental scholars of the Renaissance; but they did not let adoration of the dazzling past blind them in both eyes. Where there were glaring defects in Euclid's reasoning, the Moslems saw some of them and attempted to remedy the most conspicuous. The men of the Renaissance either noticed nothing amiss or decided to maintain a reverent silence. Consequently Euclid's geometry after 1570, the year of Dee's premeditated praise, became an article of intellectual faith as sacred as the doctrine of the holy trinity, and to question the perfection of the *Elements* was almost as dangerous for the doubter as blasphemy.

Our second popular upholder of the Platonic version of Pythagoreanism in the Renaissance is Robert Recorde (1510?-1558), physician to King Edward VI and Queen Mary, famous as the author of the first mathematical classics to be written in the English language, and the first really able exponent of the undoubted practical value of commercial arithmetic. Recorde spent his declining years in jail, probably for debt. His most illuminating declaration of faith is in his *Whetstone of Witte* (1557), in which he expounded the virtues of algebra.

"I may truly say," he confesses in an ingenious paradox, "that if any imperfection be in number it is because . . . number can scarcely number the commodities of itself. . . . As number is infinite, so are its commodities. This number also hath other prerogatives, above all natural things, for neither is there certainty in anything without it, nor good argument where it wanteth. Plato and Aristotle search all secret knowledge and mysteries by it [as in the *Timaeus*, with what results we have seen]—not only the constitution of the whole world is referred to number, but also the constitution of man [we saw that, too], yea, and the very substance of the soul [that also] . . . Beside the mathematical arts there is no infallible knowledge, except it be borrowed from them."

Until learned and influential men either stopped repeating these antique exhortations to worship at the shrine of number, or were ignored by less learned men who were not too Greek to get their hands soiled occasionally, science as we know it did not exist even in embryo. If there was any point in preaching the gospel of number as the key to all the sciences while never bothering to fit the key to any lock, what that point may have been is now obscure.

Recorde died in 1558. Only about half a century was to elapse before talk was forgotten in action. A century later the men who used numbers in their scientific work had very little to say about them. They were so busy exploring "the constitution of the whole world" (or at least the constitution of a manageable fraction of it) that they had neither time nor thought to squander in idle praises of their useful implements. Yet it has been argued that such scientifically empty Renaissance philosophizing as the specimens just exhibited "prepared the way" for Galileo and modern science. What way? Galileo was no numerologist, nor were Newton, the Bernoullis, Euler, Lagrange, and Laplace scientific mystics. Not till scientists departed from the mystical way of numbers by which, as Recorde recalls, Plato and Aristotle—not to mention Pythagoras—sought "all secret knowledge," did they come face to face with nature. The way prepared by the Renaissance eulogists of numbers and of mathematics in general was not followed by modern science, but by the resurrected Pythagoreanism of the twentieth century.

In the amber morning twilight of the sixteenth of February in the year of grace 1600, two shadowy figures stood conversing on the crest of the seventh of Rome's eternal hills. Their voices were low, for they had just emerged from the eighth—counting from the bottom—of Dante's infernal circles. After a brief silence during which they breathed deeply of the clean, cold air, Pythagoras turned to his friend.

"We met some pretty decent people down there."

"Yes," Plato agreed. "And some quite intelligent ones, too. That fellow in the big red hat we saw in Malebolge, for instance. What was his name again?"

"You mean the one who was trying to talk himself out by saying it was all a mistake that he ever got in?"

"That's the one. He said he was a bishop or a cardinal or something. It made no sense to me."

"He claimed to be both," Pythagoras reminded him.

"So he did," Plato admitted. "That's what made it all so confusing. How can one be two, even in purgatory? I must ask him to explain it to me, if he ever gets out. I wish I could recall his name in case we do meet again. What in Hades was it?"

Pythagoras remembered. "Nicholas. Son of a poor fisherman. Father's name Krebs, meaning crab. Came from Cusa, or some such place. Confirmed hater of Aristotle."

"Ah." Plato's face brightened. "That's why he made such an impression on me. Do you suppose he'll find his way up, now that we're out?"

"He won't try. He's having too good a time pestering the people who disagreed with him. Better for him to stay where he is."

Plato looked thoughtful. "Better for them, certainly. I seem to remember him saying he would have to stay till Aristotle arrives. I'm glad we got out before he does. We've that to be thankful for, anyway. Well, the last of it wasn't too bad, was it?"

"Not for you," Pythagoras admitted somewhat grudgingly. "After the grammarians adopted you, things were easier." The master made a sour face. "But I do think," he protested, "the translators might have given me and not you the greater part of the credit."

"But I did my best to tell them it was you—" Plato began, only to be cut off.

"Just as you did with Socrates. Putting off all that arithmetical metaphysics about ideas being numbers on me. I never meant any such thing. And all that human anatomy and physics and astronomy in your *Timaeus* you fastened on me and made those dumb gabies down there believe. Just like a philosopher."

"Come, come now," Plato soothed. "Don't let us quarrel on a beautiful morning like this—our first hour in the pure upper air. Are you not my friend, my other self?"

"I hope not," Pythagoras muttered. "Your time's all back end forward." But Plato hurried on as if he had not heard.

"And did not I, as your other self, strive to get your greatest discovery accepted as it deserved? And didn't I succeed—gloriously? Why, we even heard rumors of my success—yours, I mean, in Reality—just before we got out. They told us everybody who amounts to anything up here is saying the universe is a construct of geometrical units bound together by the harmonies of numbers. I shouldn't be surprised, if we get about a bit, to find all the important men going on where you left off. They know now that I was right—I mean you were right—when I said—when you said, I mean—that the only way to understand nature is by doing mathematics and nothing else."

Pythagoras was mollified. "There were rumors to that effect." A bird flew by, and a swift shadow momentarily darkened the master's brow. "Somehow," he confessed, "I feel less hopeful than you. All that clamor just before we got out about this upstart Galileo was none too reassuring, even if they are preparing for him down there."

"But Galileo is a mathematician," Plato protested. "If what he is reported to have said is correctly rendered, I shouldn't mind having said it myself."

"What, for instance?"

"Don't you remember what that fat old fellow with three hats on his head told us as he passed us on his way down?

It went something like this: 'Philosophy is written in the great book of nature, continually open for us to read. But only he who has the key to its cipher can read the book. The key is mathematics. Yet only the One, the Eternal Geometer, knows and can decipher all the diagrams in the book and perceive the everlasting truths uniting them in one supreme Truth. What the mathematician reads is but a word or two, or at most a line, and that he painfully spells out a letter at a time only after much thought. But though little, what truth his mathematics gives him is all he knows. The rest is opinion.' So Galileo is reported to have said. Be reassured, my friend of many hells, he is of the true faith."

"Then why does he always keep messing about with his hands?" Pythagoras demanded with a suspicious glance at his friend.

Plato sighed. "A youthful indiscretion. He will outgrow it. You did, you know."

"Youthful? Why, he's thirty-six if he's a day."

"Give him time."

"I'll give him time but not eternity. How long do you think this indiscretion of his is going to continue?"

Plato made a rapid mental calculation. "About three hundred years."

"But no man since Eber begat Peleg was ever condemned to one turn in the flesh that long. You mean he is to be bound to the Wheel for six or seven whirls after this, before he is pure enough to escape?"

"Roughly that. He will pass through several reincarnations —the blind will call him 'his followers.' But I give him only three centuries more of it."

"Why that?"

"Oh, it's just an eternal necessity of Ideal Numbers. Anything human that's fairly new runs its course in about that span. It's the universal three that does it—begetting time on the decadic decad. That gives three hundred years. You had about that long at your best. So did I. Then the Wheel lurched and turned up—"

"Yes, I know. Please don't remind me of it now. I'm still a bit queasy. So you and I are to be out of it for three hundred years?"

"More or less. Numbers aren't what they were in my day, and I may have made a slight error. To be safe, let's add four pentads. That's both just and generous. You agree?"

"I must. How can I deny my own numbers? So we don't come back—in Reality—till 1920?"

"That's what I make it."

"Then let us say farewell here, on this bare hilltop overlooking the city these deluded men call eternal. Both of us are tired and need rest. Let us sleep, but not too soundly, lest they forget us forever. And when we wake refreshed we shall meet in—"

The master abruptly interrupted himself. "What are all those men doing down there in the city square?"

Plato peered through the clear morning air. "It looks like preparations for a funeral. They're piling faggots. Round a pole stuck in the ground. It's all different from what it was in my day. When we cremated a hero, we ranged the logs in an orderly rectangular parallelepipedon—not all heaped up like a bestial participation in the ideal circle." He followed the business in the square with puzzled attention. "From the size of the pyre the hero must have been a prominent citizen of the place. They said nothing down below about any notable man dying recently. I wonder who he was?"

Pythagoras chuckled. "Aristotle. They're going to cremate what's left of him."

"Don't jest about fatal things," Plato reprimanded him severely. "Aristotle is not quite dead even yet."

The master apologized. "It was unseemly of me. Still, Aristotle has no right outliving us, even if he is almost dead. Staying to see the show?"

"No. There's nothing to be done here. I'm off."

"Where to?"

"The Absolute knows. And you?"

"The same. Glad to have met you, even in purgatory. See you again in 1920. Farewell!"

"Truth be with you till two are one again. Farewell!"

CHAPTER - - - - - - - - - - - - - 22

Saints and Heretics

PYTHAGORAS escaped from purgatory only when his tormentors ceased being taken seriously by men capable of independent thought. Two major tactical blunders by the opponents of free thought in the Renaissance did much to liberate science and dismiss to the past the unaided reason as the infallible guide to "reality" and "truth." In their dealings with Giordano Bruno (1548-1600) and Galileo Galilei (1564-1642), the guardians of orthodoxy overstepped the limit of human decency, and in so doing earned the efficient enmity of all free minds. Though it has tortured and hanged or burned all heretics to its own creed on whom it could lay its hands, no tyranny yet has succeeded in destroying all opposition to itself. Always one or two have escaped the stake or the noose to infect thousands with their heresies, until the tyranny died, not of the hatred which it had well earned, but of contempt.

In their fight against bigotry and ignorance, Bruno and Galileo were only two of many who, like Galileo, recanted, or who, like Bruno, burned. They have been singled out here because certain aspects of their work have an oblique but important bearing on the Pythagoreanism and scientific

Platonism of the twentieth century. Each in some measure anticipated modern speculations on the infinite.

We observed that Anaximander is credited with the first recorded hints of a philosophy of the infinite. We noted also how Zeno's paradoxes have affected nearly all attempts from his day to the present to construct a consistent arithmetic of "infinite numbers." In the Middle Ages an enormous amount of scholastic logic was expended on the theory of the infinite as it applies to Christian theology. Much of what the scholastics accomplished is parodied, not too unfairly, in the taunting jest, "How many angels can stand on the point of a needle?" While it would be fatuous to kill a joke by trying to explain it, we must note that the problem is less silly than it sounds. Transposed to the language of infinitesimals in terms of which both Kepler and Newton thought occasionally, with great profit to the physical sciences and astronomy, the problem is no more ridiculous than many that engaged the serious attention of early workers in the differential and integral calculus. Though the scholastics argued in terms of theological technicalities, while the mathematicians of the Newtonian age preferred to conduct their wrangles in the early language of the calculus, all were talking about essentially the same thing.

If the scholastics failed to reach a satisfactory theory of the theological infinite, some of them at least recognized that Aristotelian logic is inadequate for the task. Their failure was no greater, relatively, than that of the mathematicians, who have yet to produce a theory of the mathematical infinite acceptable to a majority of competent experts. Until such a theory is forthcoming, the traditional claim that mathematical reasoning is more reliable than any other, and that the truths attained by such reasoning have a greater validity than those reached by other means, is not wholly justified. These statements of what many competent mathematicians hold to be obvious facts are not intended as aspersions on the acknowledged power and the innumerable triumphs of mathematics in its own right and in the domain of the sciences. They are recalled merely that we, in our turn, may not emulate the bigotries which failed to silence Bruno and Galileo. And finally, if it seems irrelevant to cite the angels standing on the point of a needle as witnesses to the truth of mathematics, we may remember that Georg Cantor (1845-1918), founder of the modern theory of the mathematical infinite, was a close student of mediaeval theology. Of course Cantor's devotion to the logic of the scholastics may only have been one of his many aberrations. If so, it might be the cause of the extensive revisions of his theory which his successors have found advisable.

To give Bruno's capital heresy its proper setting, and the one which he himself would have desired, we must borrow a few details from the work of a most remarkable man who died a hundred years before Galileo was born. "The divine Cusanus," as Bruno called Nicholas of Cusa (1401-1464), was the son of a poor fisherman. By his own ability and his rugged independence of mind and character he hacked out a distinguished career for himself in ecclesiastical politics. As a reward for siding with the Pope in a dispute over who was to govern whom, Nicholas was appointed bishop of Brixen in the Tyrol and was awarded a cardinal's hat. His passion was mathematics.

Having the twofold advantage of humble birth and a good

head, Nicholas recognized high-flown nonsense when it was thrust at him. The excessive refinements of the Aristotelian logic and natural philosophy of his contemporaries provoked his rage, and he proceeded to take its practioners apart. Pure reasoning alone—as practiced by the hair-splitters—could never, he said, get anywhere in understanding either nature or the deity. This, of course, was heresy. But Nicholas was a power in the Church, and he suffered no inconvenience. He insisted that reasoning must be supplemented by observation and experiment. This was disgusting; but those who did not like it had to swallow it nevertheless—as long as Nicholas was about. The reasoning which should accompany empirical science, the hatted heretic preached to his squirming congregations, was not Aristotelian logic but mathematics. A century and a half later Galileo came to the same conclusion.

In his own mathematical philosophy Nicholas was a Platonic Pythagorean. He could scarcely have been anything else in his mystical day. His applications of mathematics to the grosser aspects of theology were prudent, as befitted the fisherman's son, but hardly successful. Among other notable exploits he calculated the date of the end of the world. Shrewder than the majority of those who had done the like before him, Nicholas did not make the embarrassing mistake of getting an answer which would find him alive and hearty on a Judgment Day that had been indefinitely postponed. His figuring gave him a date in 1734 as the Day of Wrath, safe by a margin of 233 years even if he should live to be a hundred. Equally happy in his attempt to square the circle, he succeeded in convincing himself that he had done the impossible. This disaster, inevitable at the time, was more than compensated by an acute anticipation of certain details of the Copernican astronomy. From all of this it is clear that Nicholas was a more fortunate unfortunate than Bacon. Though far ahead of his time he did not suffer for his lead over the learned rabble.

Probably unaware of the far-reaching implications of his main philosophical heresy, Nicholas in his most individual work took up an advanced position on the provenance of Aristotelian logic. His contemporaries ascribed this logic to the deity. Or, if that is putting their case too favorably, they at least believed that Aristotle's logic is an eternal necessity divinely imposed on the human mind by "the structure of reality." Moreover they reasoned as if they believed that even the deity must submit to Aristotle's "A is A; everything is either A or not-A; nothing is both A and not-A"—as the three fundamental laws (or postulates) of Aristotelian logic are sometimes stated. The second, called "the law of the excluded middle," is the one of present interest.

According to this law itself, the laws of classical logic either were revealed to Aristotle by the deity, or they were not revealed to Aristotle by the deity. There is no middle ground. Agreeing with some of the more progressive modern psychologists, Nicholas favored the second alternative. How, then, did the three laws originate? If the hypothesis of divine revelation is rejected as unverifiable, there is the possibility that Aristotle abstracted his laws from everyday sensory experience, precisely as Thales abstracted his lines as lengths without breadth. All the collections of things with which Aristotle (or anyone else) was (is) acquainted in that experience were (are) finite. Obviously then, according to Nicholas, Aristotelian logic is not adaptable to the true theology. The deity, by hypothesis, is infinite, not finite. Nicholas was not quite original enough to imagine the possibility that the deity may be finite but unbounded; mathematics had not got that far in his day. But he did see that the law of the excluded middle is not safely applicable to the infinite, thus anticipating one school of twentieth-century mathematical logicians.

Aristotelian logic being impotent to compass the infinite, Nicholas argued in favor of a mystical approach to the perfection of the deity. Though man is finite and can never attain the infinite, he may apprehend its existence through "mathematics, the only truth of science." The infinite toward which man strives projects its clearest image in the endless sequence of natural numbers 1, 2, 3, . . . , each of which, after the first, is generated from its immediate predecessor by addition of the unit. Here surely was a symbol of the creation of all things by the author of the universe. Nicholas had advanced a step beyond Pythagoras: the numbers were not the things their generation from the Monad symbolized, but merely an image, comprehensible to a finite mind, of the reality known only to the deity. It followed that man knows only appearances, never reality. However, human beings are not thereby doomed to eternal ignorance: at least a phantasm of reality may be visioned through that pure symbolism which is mathematics.

It was but natural that Nicholas should proceed from this mystical conclusion to an impassioned plea for the application of mathematics to the understanding of nature. The mathematics of his time was inadequate for such a program on any significant scale. When the necessary mathematics was devised by Newton in the seventeenth century, it made bold use of the mathematical infinite and none at all of the theological infinite advocated by Nicholas. Fortunately for his peace of mind, Nicholas had been dead almost exactly two hundred years when Newton invented his calculus and began applying it to dynamics and astronomy.

Nicholas died at the age of sixty-three, full of good works, honors, and heresies. He was entombed with the double pomp due by right to a great bishop and a wise cardinal. He has not yet been canonized.

Bruno (1548-1600) went much farther than his "divine Cusanus." Nicholas had merely disliked the Aristotelians. Bruno hated them. For any Italian in the sixteenth century such an attitude was an invitation to persecution, as Bruno well knew. Still he would not or could not abate his truculent ridicule of Aristotelianism and all its works including, unfortunately for him, the logic by which the theologians had secured the official religion. Nor could he restrain his enthusiasm for the Copernican astronomy, a heresy only less damnable than outright denial of the divine inspiration of the holy scriptures. To leave no possible doubt in the minds of his adversaries, Bruno embraced that heresy too, classifying the sacred miracles and inspired teachings with the myths and superstitions of primitive peoples. His own substitute for what he scornfully called the delusions of his learned contemporaries was a poetic pantheism. In it he combined the teachings of the "divine Cusanus," fragments of Neo-Platonism, the essence of Pythagoreanism, Copernican astronomy, and scraps from the Stoics and the Epicureans with cosmic speculations of his own. All this was compacted into one colossal heresy that contradicted in general and in detail everything sacred to the Aristotelians and theologians of all denominations. With all these eccentricities to his credit, it seems rather remarkable that Bruno should have been a devout numerologist. He had much to say on the Pythagorean decad, and a great deal on the number five that no disciple of the master could possibly have imagined.

If Bruno's system can be said to have had a unifying thought, it was his idealized version of the Pythagorean Monad—the One, the unifying principle from which all things come and to which all things return. Reminiscent of Anaximander's infinite, Bruno's Monad was his pagan substitute for the deity sanctioned by the Aristotelian theologians. His higher numerology was refined, but made no significant advance beyond Plato's: Ideas are generated from the One. More significantly, the universe for Bruno was infinite.

A particularly disturbing effect of Bruno's comprehensive heresy was its repercussion on Dante's *Comedy*. Paradoxically, Bruno's infinite cosmos provided both too much room and too little for Dante's paradise and left no space at all for his inferno. Where, exactly, were these celestial and infernal regions, so graphically and so minutely described by the poet, located? According to Bruno, nowhere. But the teachings of the constituted authorities had embedded these poetic fantasies so deeply in the popular consciousness that the contours of hell were better known to literate and illiterate alike than the familiar hills and dales of the countryside. And lest some light-minded peasant should forget the way to Malebolge on the sixth day of the week, he was reminded on the seventh by detailed masterpieces on the walls of his place of worship. These alone would probably have sufficed to keep his feet from straying. The accompanying exposition by his spiritual adviser of what awaited him beyond the grave, should he err from the path of duty and obedience to those in authority, was terrifying but largely superfluous.

Bruno brushed aside all this secular cosmography of celestial love and infernal hate. Copernicus had shattered all but the outermost of the celestial spheres of the Aristotelians; Bruno smashed the Dantean heaven and hell of the theologians. Between them they abolished the tight little egg into which the ancients had cramped the universe, and freed the mind to think and explore endless space. This was the unpardonable heresy. If, as Bruno said, there was not one world but an infinity of worlds, then the atonement and crucifixion must be enacted infinitely often to save the souls of the inhabitants of those worlds. The logic was sound. Bruno must be silenced. But how to catch him?

Skillfully tempering outspoken rashness with alert wariness, Bruno succeeded in eluding his enemies until he was betrayed by his friends. Educated as a Dominican friar, the universal heretic imagined when skepticism first assailed him that the Calvinists of Geneva would welcome him to their own disobedient fold and offer him shelter. But the Calvinists still believed in much that Bruno derided as superstition. They invited their plain-speaking guest to leave. He fled to Paris, there to land squarely in the most active nest of Aristotelians in all Europe. Nothing about them pleased the contumacious freethinker. Crossing the Channel, he proceeded to Oxford. There a modicum of English common sense made the scholarly atmosphere almost endurable, and Bruno lectured freely on Copernican astronomy and his personal heresies. But the restless man could not resign his soul to comparative tranquillity, and he returned to the Continent.

He was now notorious. One center of learning moved him on to the next. Some he left of his own will when he could no longer tolerate the bigotry and stupidity of his colleagues. Wittenberg, Prague, Helmstedt, Frankfort, Padua, and finally Venice—all sheltered him for a time while Rome, with infinite patience, sought to trap him. The friendliness of the Venetians dulled his caution. They were sympathetic and seemed really to like him. But their sympathy exceeded their courage; and when the long arm of the Holy Inquisition reached for the heretic of all heretics, they gave him up.

The Inquisition was not hasty. It granted Bruno seven unpleasant years in which to recant and repent. Obstinately defiant, he refused to do either. Their patience exhausted at last, the guardians of orthodoxy decided it was time to turn the clock back a thousand years. For the good of his immortal soul Bruno was burned at the stake as an incorrigible heretic on the sixteenth of February in the year 1600.

Two unforeseen consequences of that particular bonfire astounded and disconcerted those who had applied the torch. The fire got out of control. It flared up in a blaze that consumed the last rotten remnants of the Middle Ages and dispelled the darkness before the dawn of modern science. In that sudden light the sadistic pyromaniacs responsible for the fire shriveled into grotesque caricatures of themselves. Their night was ended; and though they still might be feared by some, they commanded the respect of none.

To commemorate this decisive rupture between the old world and the new, Bruno's admirers in 1889 erected a statue to him on the spot where he had burned, but only after fierce opposition. It takes a tolerant mind to forget its mistakes.

After Bruno, Galileo. Bruno's failure to convert his persecutors to the Copernican astronomy was overbalanced by his total conversion of Galileo, who absorbed the great heresy from Bruno's writings.

Galileo's career is so well known that only a few relevant details need be recalled here. Though serious enough about serious matters, Galileo was no fanatic like Bruno for what he considered the right. His gift for sarcasm and satire made him a far more dangerous opponent than any deadly zealous candidate for martyrdom. The logician who challenged Galileo usually regretted the encounter. Galileo despised rather than hated the Aristotelians, and his acid contempt stung more deeply than Bruno's earnest bludgeoning. He also had the exasperating advantage of being a devout believer in the official religion of his day. Though there never has been any question of the sincerity of his protestations of faith, there can be no doubt that Galileo's orthodoxy in this respect served him as well as any other protective coloration that nature or his own wit might have devised. Bruno invited persecution; Galileo adroitly avoided it. But at last the bigots he had ridiculed caught up with him.

A sharp warning in 1616 had cautioned Galileo to moderate his enthusiasm for the new astronomy. He complied, more or less. But he was not the man to stultify himself to increase the prestige of anybody. In his great dialogue (1632) on the Ptolemaic and Copernican systems of astronomy, he unequivocally championed the latter. But the Aristotelian ecclesiastics had decreed that the former is the true astronomy. Numerous explicit statements in holy writ disagreed with Copernicus and therefore also with Galileo.

The prolonged game of cat and mouse, with the man of science in the role of mouse, came to an end. Galileo was haled before the Holy Inquisition. There followed a trial that still is a classic of tediousness and fatuity. The heretical astronomer was condemned (June 22, 1633) to abjure the Copernican theory and his own teachings as contrary to holy writ.

The official document sentencing the culprit to abjuration, technical life imprisonment, and the recitation of the seven penitential psalms once a week, was signed by seven of the ten cardinals who tried him. Galileo recanted. He was in his seventieth year, broken in health, and he had been humiliated before fools. With admirable good sense he declined the opportunity to participate in another Roman holiday such as Bruno had provided.

Mathematicians should be interested in reading the original document—too long for reproduction here—as in it, for the last time in history, they and their practices were singled out for especially severe official censure. Since that historic disapproval mathematicians appear to have been too insignificant for any official's condescension.

Galileo's part in the inauguration of modern science is sometimes minimized by historians of science, but never by working scientists who know something of the history of science. It is true that others had talked of combining mathematics with observation and experiment. Nor was Galileo the first to insist that the principles of natural science should be obtained empirically, stated mathematically where possible, and be made the basis of a deductive system whose conclusions can be tested empirically. But he said it more clearly and more explicitly than others. What is more important, he was the first to supplement eloquence with action on a scale that showed all but the wilfully blind that the method he advocated, and practiced, succeeded where others failed.

Of Galileo's contemporaries and rivals for fame, Descartes

(1596-1640), often called the first modern philosopher, was one of several who said almost as much as Galileo about scientific method. But the philosopher's genius was essentially mathematical and abstract; and beyond an ill-concealed jealousy of Galileo, Descartes paid no attention to him as a scientist. We have seen that Descartes was a Platonic realist in his mathematics, and that Galileo praised mathematics as roundly as Plato himself. But whereas Descartes was content with his mathematical realism, Galileo did not remain constantly in an attitude of adoration. He went to work.

Galileo's particular interest for us here is in his contribution to the theory of the mathematical infinite. From its satirical setting in one of his dialogues it is impossible to judge whether Galileo himself took his epochal remark seriously, or whether he just tossed it off maliciously to confound a stupid Aristotelian in his own logic. Whatever his motive, Galileo isolated the capital distinction between finite and infinite collections.

By "part" we shall mean some but not all. In a finite collection there are always more things in the whole collection than in any one of its parts. Galileo showed by an example that a part of an infinite collection may contain the same number of things as there are in the whole collection. Two collections are defined to contain "the same number" of things when the things in them can be paired off, one from each collection in each pair, in such a manner that the pairing leaves no thing in either collection unpaired. This is merely an explication of what we mean, subconsciously perhaps, in common speech when we say that two collections contain the same number of things.

Examples of collections in which a part contains as many things as the whole are easily exhibited. All the even numbers 2, 4, 6, 8, 10, . . . are only a part of all the natural numbers, 1, 2, 3, 4, 5, 6, 7, 8, 9, 10, . . ., yet there are the same number of even numbers as there are of all natural numbers. The pairing is effected by putting each natural number with its double:

$$\begin{array}{l} 1, 2, 3, 4, 5, \ldots \\ 2, 4, 6, 8, 10, \ldots \end{array}$$

Galileo's example,

$$\begin{array}{l} 1, 2, 3, 4, 5, \ldots \\ 1^2, 2^2, 3^2, 4^2, 5^2, \ldots, \end{array}$$

in which each natural number is paired with its square, is even more spectacular.

This ingenious observation was the first hint that it may be possible to talk consistent mathematics about "the infinite." It appears to have been effectively overlooked by mathematicians till the early nineteenth century, when others noticed the seeming paradox about "whole" and "part" with reference to collections that are not finite, accepted it as a fact, and began serious work on the mathematical infinite. By the end of the century there was a highly developed theory of the infinite as it occurs in both pure and applied mathematics, including an arithmetic of infinite numbers. As noted in an earlier chapter, the subtle paradoxes that slipped into some of this work occasioned a closer scrutiny of all deductive reasoning than any since the time of Aristotle. This in turn cast suspicion on the status of mathematics and logic as revelations of eternal truth and divine necessity.

Had the ten cardinals who tried Galileo been aware of his thoughts on the infinite they would have wasted no time on his Copernican heresies. His provocative example of all the natural numbers and their squares was to disturb the logic on which the mediaeval authorities had based the official theology more profoundly than all the unorthodoxies of the new astronomy. But it is difficult to see how Galileo could have recanted on that example.

Rhetorical laudation of mathematics as the divinely inspired answer to all the riddles of the universe went out of fashion among active men of science with Galileo. The antiquated anthems continued to pour forth in undiminished volume, it is true, but not from those who were creating the new mathematics and applying it to the physical sciences and astronomy. Whatever mathematical mysticisms the leaders may have cherished were strictly off the record. Not till the late 1920's did a few working scientists pick up the chant in praise of "divine mathematics" where Galileo had dropped it. But progress—or retrogression—was rapid, and by 1930 the Platonic deity returned after an absence of three centuries as the Great Mathematician. Simultaneously the universe became a mathematical thought, like a complicated geometrical theorem, in the same Mathematician's mind.

The decline of rhapsodic mathematics among practicing scientists seems to have been largely occasioned by the hard common sense of Newton. Born in 1642, the year of Galileo's death, Newton lived through the first quarter of the eighteenth century, dying in 1727. His *Mathematical Principles of Natural Philosophy* (*Philosophiae Naturalis Principia Mathematica*) of 1687 became the scientific bible of the great Continental mathematical astronomers and mathematical physicists. Only in a scientifically superfluous appendage to the second edition of the *Principia* is there a suspicion of anything that might be called mathematical mysticism. Criticized by Leibniz (1646-1716) and by Bishop Berkeley (whom we shall meet in another connection) for having ignored theological metaphysics in the first edition (1687) of the *Principia*, Newton added a "General Scholium" on such matters to the second edition of 1713. If Newton's orthodox admirers had expected a flattering endorsement of their own beliefs by the master mathematician and scientist of the age, they must have been somewhat disappointed with Newton's specifications for a supreme being. All anthropomorphism was repudiated. What remained was like nothing Newton's devout contemporaries had hoped he must imagine.

Newton's conception of the deity is of some mathematical interest for its repeated insistence on the infinite as a characteristic attribute of the supreme being. The deity, according to Newton, "is supreme, or most perfect. He is eternal and infinite, omnipotent and omniscient; that is, his duration reaches from eternity to eternity; his presence from infinity to infinity. . . . He is not eternity and infinity, but eternal and infinite; he is not duration or space, but he endures and is present. He endures forever and is everywhere present; and, by existing always and everywhere, he constitutes duration and space. . . . He is utterly void of all body and bodily figure. . . ."

It would be interesting to know what Galileo's decad of cardinals would have thought of all this. Perhaps some of them read it in a better world. But there they could do nothing to suppress it or to silence its author. Thanks partly to

Henry the Eighth and his numerous wives, the Holy Inquisition had no active agent in England, and Newton was free to believe what he pleased and to publish his belief if he wished.

Like the sound Englishman he was, Newton had hoped—as he told a friend—that his mathematical astronomy would supply his contemporaries with a rational conception of the deity. Newton's British successors frequently concluded their own scientific expositions with comments rather less provocative than his on the theological implications of their researches, or at least on a note of humble praise. This pious proclivity of their British colleagues never failed to amuse the lighter-minded Continentals. It went out of fashion about the middle of the nineteenth century.

Though there may be an aura of mathematical mysticism about Newton's deity, there is no taint of numerology in either his theology or his science. Temperamentally Newton was a modern Thales in his possession of a forthright common sense. The greatest of natural philosophers, he did not permit his metaphysics to hold him back when he wished to go forward. The absolute space, absolute time, and absolute motion of the *Principia* might have been clear to Plato. They were not clear to Newton. But he saw that, even if clarified, these obscure absolutes were irrelevant for the matter in hand, and he went ahead without wasting time over them. They had no more importance for his purpose than his remarks on the nature of the deity. Likewise mathematics was subordinated to the main purpose. Consequently mathematical mysticism was temporarily ousted from reputable scientific thought with the *Principia*.

But in professional philosophic speculation the old magic

of numbers lingered on, as fantastic as ever. Leibniz (1646-1726), the leading philosopher of his age and one of the few universal minds in history, observed that 1 and 0 are the only digits in the binary scale of notation. From this he inferred that the deity (1, the Monad) had created the universe out of nothing (0, zero). Yet this vestigal Pythagorean invented the calculus independently of Newton and only about twenty years later. Leibniz was never known to joke. We are in the presence of a miracle.

The next critical episode in the progress of mathematical mysticism concerned the singular mishap of Euclid's unquestioning worshipper and unwitting critic.

CHAPTER - - - - - - - - - - - - - - 23

A Turning Point

THE year 1733—six years after the death of Newton, and just a year before the end of the world predicted by "the divine Cusanus"—marks a definite turning point in the career of mathematical mysticism. Pythagoreanism and Platonism in science and mathematics might have been abandoned then, had scientists and mathematicians been aware of the work of Girolamo Saccheri. But so far as recognition went, Saccheri was another Roger Bacon. The crucial detail was the modified conception of geometric truth which might have followed immediately after Saccheri, but which was delayed for nearly a century. It concerned the status of Euclid's geometry.

Euclid's *Elements* are said to have passed through more printed editions than any other book except the Bible. Probably more directly than any other one mathematical work, the *Elements* were responsible for perpetuating the creed that "mathematical reality lies outside us." For generation after generation hundreds of thousands, if not millions, of docile students of elementary geometry were convinced by the rigidity of Euclid's propositions that his was the one thinkable account of space. Only in 1903 was "Euclid" universally abandoned as the textbook for schoolboys which it was never intended to be. Its last pedagogical refuge was in the secondary schools of England. An obstinate fight of about thirty years finally got it out, and "Euclid" as a synonym for school geometry at last became a dead word in all civilized languages.

Euclidean geometry—not Euclid's *Elements*, but the type of geometry taught in the usual school course—is still the simplest and most useful of all geometries for everyday life and for by far the greater part of physical science. But workaday utility is not the only thing of practical importance for our generation. What our predecessors inferred from the tactics of geometrical proofs in their struggle to give meanings to "truth" and "reality" is equally important. The impact of elementary geometry on their habits of thought was as practical for them, and through them for us, as were all the machines ever designed in accordance with Euclid's geometry and Newton's mechanics. The absolutism of geometric "truth," drilled into adolescents in their formative years, conditioned educated but uncritical minds to accept absolutism in the "truths" of other intangibles from philosophy and religion to economics and politics.

Before following the decline of Euclidean absolutism we may recall the little that is known about its immortal author. So deeply was Euclid submerged in his work that almost nothing of the man himself has survived. His dates are uncertain, but 330-275 B.C., frequently given, are probably close enough. It is supposed that he was educated in Athens, perhaps in the Academy. Attempts to make Plato's estimate of mathematical truth responsible for the form of Euclid's *Elements* are based wholly on unsubstantiated conjecture. The *Elements* was composed at Alexandria, where Euclid spent most of his life as a member of the scientific staff centered at the great library.

The book is a systematic compilation of the elementary geometrical and arithmetical knowledge of its time. Euclid's personal contribution was the arrangement of all this scattered material into a logical sequence, in which everything was supposed to be derived from explicitly stated assumptions by the accepted rules of deductive reasoning. The measure of his success in this ambitious project is the detail of capital historical interest. In the geometry it was negligible.

If this seems too harsh, any impartial critic may convince himself in less than an hour—as many did when European geometers began to recover from their uncritical reverence for the Greek mathematical classics—that several of Euclid's definitions are inadequate; that he frequently relies on tacit assumptions in addition to the postulates to which he imagined he had restricted himself; that some of his propositions, as he states them, are false, and that the supposed proofs of others are nonsense.

The attempted proof of the very first proposition of all —"on a given finite straight line to construct an equilateral triangle"—is erroneous beyond repair. It is vitiated by a glaring oversight which any schoolboy who has been cautioned to use his brains as well as his eyes will detect immediately. It is impossible to patch up Euclid's attempt by using only the assumptions which he permitted himself. His proof of the second proposition depends upon that of the first; so it also is fallacious. The third cites the second; so it too collapses. And so on, to the seventh, which is meaningless. If it were worth anyone's trouble the entire logical structure of the geometrical portions of the *Elements* might be destructively analyzed for inexplicit assumptions and defective proofs. Yet the thirteen books of Euclid's *Elements* for over two thousand years were worshipped by the uncritical as the embodiment of logical perfection. Though our own attempts at precision in reasoning may suffer the same fate as Euclid's, his example has taught some mathematicians caution in their claims for the eternal truth of their personal contributions.

An interesting conjecture as to Euclid's motive in composing his masterpiece takes us back to the Pythagoreans and Plato. The first proposition essays to construct the simplest regular polygon (an equilateral triangle). The concluding six propositions of Book XIII, the crown of the entire work, give constructions for the five regular solids. Thus the geometrical parts of the *Elements* might be regarded as the mathematical framework of the Pythagorean cosmos as elaborated by Plato. Euclid's alleged purpose was to secure this framework against all rational doubt.

Four of the critical dates in the evolution of geometric truth are 1701, 1733, 1781, and 1826. The names connected with these are respectively George Berkeley (1685-1753), Girolamo Saccheri (1667-1733), Immanuel Kant (1724-1804), and Nicolai Lobachewsky (1793-1856). Berkeley was an Irish metaphysician, theologian, and finally a bishop; Saccheri, an Italian Jesuit logician, theologian, and mathematician; Kant, a German philosopher of Scotch descent and no mathematician; Lobachewsky, a Russian mathematician and no philosopher. With the exception of Kant each of these men made a definite contribution to the de-Platonizing of mathematics in general and of geometry in particular.

Berkeley seems to have been the first metaphysician on

record (with the dubious exception of some mediaeval nominalist) to suspect that there is nothing absolute about the "truth" of geometry.

Saccheri, against his professed intention and contrary to his ineradicable belief, was the first to demonstrate that Euclid's system of geometry is not the only one conceivable.

Kant simply made a complicated mistake. In his own technical sense, to be described in the proper place, he regarded the truths of elementary (Euclidean) geometry as apodictic and synthetic *a priori*. Roughly, he believed that geometrical theorems, such as "the sum of the angles of any plane triangle is equal to two right angles," are necessary truths inherent in the nature of reality as presented to thought by the very structure of the mind: the human mind can apprehend geometric truths only through the mold of Euclidean geometry. This geometry therefore is imposed on human beings by the natures of reality and the mind. It is the only one possible.

Lobachewsky, well aware of what he was doing and what his work implied for geometric "truth," produced a fully elaborated system of geometry, self-consistent and different from Euclid's and, like his, adequate for everyday life. Consciously and deliberately Lobachewsky did what Saccheri tried to convince himself was impossible, but what, in spite of his obstinate loyalty to Euclid, he partly accomplished.

The contradiction between Kant and Lobachewsky is total. Though there have been persistent attempts to show that Kant was misunderstood, and that his metaphysics can accommodate the non-Euclidean geometries, competent mathematicians and mathematical logicians agree that Kant was mistaken. "Nothing," he said, "has been more injurious to philosophy than mathematics." Certainly nothing has been more injurious to Kant's personal philosophy than his attempt to prove the unthinkableness of a geometry other than Euclid's. That geometry for him was a necessity. Partly from analogy with this supposed necessity of Euclidean geometry, partly from other considerations, Kant evolved his theory of "things in themselves"—disguised absolutes.

Saccheri's involuntary masterpiece, containing the first specimens of non-Euclidean geometries, was printed in 1733. Kant's *Critique of Pure Reason,* with its false conception of geometry, appeared in 1781. Some devotee of Kant's system might be interested in trying to imagine what Kant would have had to say about absolute geometry and absolutes in general if he had chanced upon Saccheri's book. We shall see later why the philosopher was denied that privilege.

All this is not so academic as it may seem. No lesser an authority than Thomas Mann asserted in 1941 that what civilization was really fighting for in the Second World War was The Absolute.

CHAPTER - - - - - - - - - - - - - - 24

The Skeptical Bishop

TAKING Berkeley, Saccheri, and Lobachewsky in their chronological order, we shall consider Berkeley first.

Berkeley's life (1685-1753) overlapped the first great period of the Newtonian mathematics and natural philosophy. His intellectual career, wherever it paralleled mathematics, offers an interesting example of the progression of an original and independent mind from orthodoxy to heterodoxy. But in his primary interest of Christian theology Berkeley never wavered. If his thought had any dominant motive it was his determination to rationalize the deity by what mathematicians call an existence proof. In all his mathematical work Berkeley was the crusader for what he conceived as a divinely imposed obligation. Sometimes in his effort to make the supreme being acceptable to a reasoning mind, he could be as coldly rational as any modern mathematician composing a technical monograph on the postulates of geometry or the logic of the infinite. More frequently he was as quixotic in his thinking as he was altruistic in his life.

For no reason at all, apparently, Jonathan Swift's friend Vanessa left Berkeley half her fortune. She had sat next to him once at a dinner. The money, of course, had to be expended in philanthropy, though Vanessa willed it unconditionally. Berkeley spent some of it in trying to alleviate the shocking poverty and corruption of his own Ireland.

The highlight of Berkeley's career shows him at his most idealistic. In his early forties he gave up his comfortable living as Dean of Derry and sailed with his newly wedded wife for America. His prospects were practically nil, consisting of a feeble promise from the English government of a salary that might some day materialize. He was to "found a college in the Bermudas for the Christian civilization of America." He never reached the Bermudas. By some inadvertence on the part of the navigator, the missionary of Christian civilization was shipped to Newport, Rhode Island. There he rusticated for three peaceful years, enjoying the colonial life and observing the customs of the Indians. When all prospects of the elusive salary vanished, Berkeley returned to England.

At the age of forty-nine he became Bishop of Cloyne in his native Ireland. The same year (1734) he published his boldest mathematical heresy, *The Analyst.* Its full title sufficiently describes its purpose: *The analyst: or, a discourse addressed to an infidel mathematician. Wherein it is examined whether the object, principles, and inferences of the modern analysis* [Newton's calculus] *are more distinctly conceived, or more evidently deduced, than religious mysteries and points of faith.*

The decade following *The Analyst* brought forth *Siris,* a curious compilation which Berkeley seems to have regarded as his masterpiece. He said it had cost him more thought than all his other labors combined. *Siris* is Berkeley's mature reaction to the teachings of Plato and parts of the Neo-Platonic philosophy, both of which he had long since out-idealized.

The main topic however is peculiarly Berkeley's own. He imagined that in the "resinous element" of tar-water he had discovered the curative elixir for all human ills from spiritual disorders to smallpox.

Merely from these samples of his incessant activity it is clear that Berkeley's genius was as discursive as it was original. Only a few details of all his varied output need concern us here.

Berkeley's characteristic contrariety showed itself early. Mathematics was his first love. It was also his last, though in his later years he deluded himself into believing he had left "all that" behind him with his adolescence. At the age of sixteen he composed a creditable essay on Euclid which he published three years later. The monotonous lock-step regularity of Euclid's deductive reasoning, marching from one apparently inevitable conclusion to another, hypnotized the boy. Like Plato and "all the saints and sages" from Pythagoras to Augustine, young Berkeley beheld the vision of geometry as truth eternal. Unaware of his subconscious doubt, the dazed geometer conceived the heretical notion of giving a mathematical proof of the deity's existence. Had he succeeded, as he realized when he woke from his grandiose dream, he would have demonstrated the deity's slavery to mathematics, which would thereby have been proved to be the supreme being.

The purely symbolic and formal reasoning of algebra jarred the young man out of his daze. With a penetration which would have been extraordinary in any age and was doubly so in Berkeley's, the self-appointed champion of the deity suddenly saw clear through the supposed eternal necessity of any mathematics. There was nothing in it. Mathematics, he perceived, is analytic and humanly free; it yields nothing that has not been put into it by mortal men. It is not, as Kant was later to proclaim, apodictic, *a priori*, and synthetic. It is precisely because the body of mathematics is empty of all factual content that a logically rigorous mathematics is possible. The project of proving the existence of the deity by mathematics was abandoned forthwith.

In rejecting the Platonic reality of mathematical truth, Berkeley anticipated one modern school of thought by over two hundred years. Mathematics for the formalists of the twentieth century is a meaningless game played with meaningless marks or counters according to humanly prescribed rules—the humanly invented rules of deductive logic. Mathematics in this deflated philosophy does not report on our observations of an eternally existing and extrahuman Platonic reality; it is the "doing," not the "knowing," of a game in which the rules are as humanly arbitrary as those of chess.

What does a game of chess signify for the Eternal Verities? The Platonic answer is that all possible games of chess were laid up forever in heaven ages before any human being imagined a chess board or a set of chessmen. Aeons before the human race came into existence an ideal and celestially perfect Chess existed timelessly in the realm of Being. And so for mathematics to those who believe that "mathematical reality lies outside us." The Platonic reality of chess and of mathematics somehow sounds less doubtful than the like for bridge and poker. But as Parmenides convinced Socrates, the banal must be idealized along with the sublime if Platonic realism is to mean anything.

The Platonic theory of mathematical truth failed to satisfy

young Berkeley. Mathematics for him was nothing more superhuman or profound than an extremely efficient kind of logical syntax. And there are many who would agree with him today.

Though he had cured himself of mathematical realism, Berkeley could not cure others. The eighteenth century was about as unpropitious a time as could have been chosen for the promulgation of a mathematical heresy like his. The noisy triumphs of the Newtonian mechanistic materialism made it impossible for the all but inaudible promptings of cold, skeptical reason to be heard. The Newtonian infinitesimal calculus, rapidly being improved and developed by the Continental mathematicians, had produced a mechanics of the heavens that seemed to be truly celestial in all senses humanly imaginable. This new mathematics, believers declared, surely must contain some element of absolute and eternal truth. Possibly it did; its scientific successes were undeniable. But whatever the mysterious element might ultimately prove to be, the incredulous Berkeley showed that it certainly was not logical consistency.

Berkeley brought the same skepticism to bear on Newtonian mathematics that he had applied to Euclid's geometry. Specifically, he objected to Newton's reasoning about the infinite, particularly as applied to the "infinitely small quantities" in the early form of the calculus. Newton had ignored fundamental difficulties which, as we saw, stopped the Greeks short of modern mathematics. He was aware that he had done so. Berkeley also might have ignored them, for the sake of the scientific results obtained, had it been purely a question of applied mathematics. But it was not.

It frequently happens when the main army of science makes a major advance, that a horde of unskilled camp-followers rushes into the newly won territory to pick up the leavings. It was so after Newton. Mathematical amateurs with but little technical competence, seizing on what the professionals had ignored or thrown away, began misusing it to their own ends. Mechanical or mathematical proofs of the existence of the deity, also similar proofs of his non-existence, were freely manufactured and exposed for public consumption. There were many purchasers, especially among the intelligentsia of the time, till the hastily assembled mechanistic theology began exploding in their faces. It was this spectacle which roused Berkeley's displeasure. He was moved from mild anger to devastating action by the death of a friend.

Newton's friend Halley was one of those who got out of Newton's natural philosophy a great deal more than its inventor had put into it. Halley took it upon himself to demonstrate mathematically the inconceivability of the dogmas of Christian theology to one of his victims. The man happened also to be a friend of Berkeley's. Halley succeeded in the very year that Berkeley became a bishop. Halley's convert refused Berkeley's spiritual offices on his deathbed. The outraged bishop decided that something must be done to stop the meaningless gabble about mathematics and theology. In a cold rage he wrote *The Analyst* and addressed it to "an infidel mathematician."

Berkeley's attack on the inconsistencies and the unadmitted reliance on mysticism in the mathematical reasoning of his contemporaries was unanswerable. It remained not only unanswered but unacknowledged. Bigotry had changed sides. Deserting the theologians, it took up with the mathematicians and scientists. Who was this brash Bishop, anyway, who dared to blaspheme against Newton?

It was not till the late nineteenth century that Berkeley's criticisms were admitted to have been well founded. By then however they were only of historical interest. The controversy was as dead as the "infidel mathematician" who had inspired it.

Berkeley's fate is reminiscent of Roger Bacon's. In a less assertive "age of reason" he might have gained some celebrity as a pioneer in pure mathematics. His mastery of mathematical technique was too slight for creative work in the fashionable things of his time. But in his criticisms of mathematics itself he was far ahead of his age. His initial heresy regarding the eternal necessity of geometry was echoed and amplified in three main controversies over the meaning of mathematical truth that began in the late nineteenth century, and continued with increasing confusion and heat into the twentieth. His reading of algebra as mere syntax was resumed in the 1830's, only to fade from memory again for several decades. His denial of the Platonic reality of mathematical abstractions was in the direction of modern logical positivism. And finally his criticism of the Newtonian calculus was a halfway house between Zeno and modern critical work on the foundations of mathematical analysis. But none of this aided the persistent theologian in solving his lifelong problem of proving the existence of the deity.

Mathematics having failed him, Berkeley sought other means of establishing the postulated existence. He first convinced himself that he had disproved the existence of matter. From this it seemed to him to follow that everything is spirit. Granted this, the rest of the proof was comparatively easy. Though it was not a proof by mathematics—Berkeley had shown that such is impossible—it proceeded in the manner of rigorous mathematics. Arguments for the necessity and sufficiency of each step gave the whole a professional air of strict logical rigor.

In the *Essay toward a New Theory of Vision* (1709), for example, Berkeley attempted to prove that "visual space" is purely ideal and all in the percipient's mind. Making a generous concession to Pythagoras, Berkeley accorded number the status of reality. The *Treatise on the principles of Human Knowledge* (1710) carried the argument for pure idealism much farther, denying the existence of matter and proving that mind is the only possible "substance." It is all as convincing as a book of Euclid: grant the hypotheses and take the logical consequences whether you like them or not. The outcome of it all is the Berkeleyan "Everything," the famous "To be is to be perceived"—*Esse est percipi.*

If Berkeley had been a worse mathematician and a better bishop than he was, he might have taken at least half the theorems he proved on faith. Thus instead of filling volumes with Euclidean reasoning to demonstrate the following grand proposition he might simply have believed it: "There is an omnipresent *Eternal Mind,* which knows and comprehends all things, and exhibits them to our view in such a manner, and according to such rules, as He himself hath ordained, and are by us termed the *laws of nature.*"

It is interesting to compare Berkeley's eighteenth-century confession of scientific faith with Eddington's twentieth-century declaration of unscientific independence: ". . . all the laws of nature that are usually classed as fundamental can be foreseen wholly from epistemological considerations."

To such contradictory conclusions can the philosophic spirit of mathematics, embodied in different minds two centuries apart, attain without inconveniencing itself. The thesis of one epoch is reconciled with its antithesis in another by a Hegelian synthesis, and the outcome is a more unified and broader knowledge.

CHAPTER - - - - - - - - - - - - - - 25

Believer and Disbeliever

SACCHERI (1667-1733) made no more impression than Berkeley on the obstinate "will to believe"—in William James' classic phrase—of the eighteenth century. That hard-headed century has been called the Age of Reason, which seems rather ironical when we remember how Berkeley's *Analyst* was received by the reasoners. Saccheri's failure to shake the dogmatism of his age was due partly to his own temperament, partly to the discipline under which he was compelled to labor.

If a supreme test of faith had been demanded of Saccheri he might have prayed that something be proved possible to him in order that he might disbelieve it. Not that he was by any means a skeptic or a cynic, for he was neither. He was simply a natural genius at believing what he wished to believe. Though this is the simplest explanation of his twisted career it is not the only one possible; others will suggest themselves as we follow the devious misadventures of his masterpiece.

Saccheri's brilliant success in his struggle to convince himself of the absolutism of Euclid's geometry is one of the most curious psychological paradoxes in the history of reason. Determined to believe in Euclid's system as the absolute truth, he constructed two other geometries, each self-consistent and as adequate for daily life as Euclid's. Then, by a double miracle of devotion, he disbelieved both. As there are precisely three possibilities conceivable in the particular matter concerned (to be noted presently), one of which is Euclid's geometry and the other two Saccheri's discarded pair, it followed for the resolute believer that Euclid's is the only one possible. But this is what he wished to prove. Surpassing Saul who went in search of his father's asses and came home with a kingdom, Saccheri in seeking to find Euclid alone discovered a couple of universes. But unlike Saul, he returned with what he had set out to find.

Little is known of Saccheri's life, possibly because there is not much to be known beyond the dryly formal record of successive assignments as a member of the Society of Jesus. The Jesuits seem to sink their personalities in the discipline of their Order, and Saccheri was pretty thoroughly submerged. But before he went under forever in the year (1733) of his death, he managed to toss his subversive discovery in the general direction of a completely indifferent world. His masterpiece on non-Euclidean geometry sank out of sight with him temporarily, floating to the surface only one hundred and fifty-six years after he had vanished.

The legend of his boyhood in San Remo pictures Saccheri as never without a copy of Euclid's *Elements* under his arm, even while at play. That he frequently opened the book and scanned its contents is evident from his subsequent career. Before he was twelve years old Saccheri had signed away the freedom of his mind for life. He became Euclid's devoted slave for as long as he should continue to coddle his faith to the discomfort of his reason. If this is an unjust estimate of Saccheri's intellectual life, it may be that which he himself would have wished his superiors to form.

In other respects also Saccheri was genuinely precocious. At the age of ten he was expert in mental arithmetic. By eleven he was sunk deep in philosophy and deeper in chess—at which he was better than good. At eighteen he was immersed in theology and well on his way to becoming a respectable Jesuit professor. On finishing his novitiate at the age of twenty-three (1690), he was assigned by his superiors to a Jesuit College in Milan as instructor in rhetoric, philosophy, and theology. Following various transfers to better teaching positions, he became professor of mathematics at the University of Pavia.

Saccheri was quite a prolific writer. Apart from his epochal failure to establish Euclidean absolutism his major works were in logic. The penetrating acuteness of his mind is shown also, according to competent judges, in his contributions to theology. The snarled problem of divine grace, for example, exercised all his subtle skill in casuistry before an acceptable solution was teased out.

The nature of Saccheri's greatest achievement can be seen very simply. As already implied this marks a decisive turning point in the evolution of geometric truth and in the conception of "mathematical reality." It therefore merits some consideration.

If either of two propositions implies the other, the propositions are said to be equivalent. Otherwise stated, the propositions A, B are equivalent if A implies B and B implies A. If either one of two equivalent propositions is proved, so is the other.

Now Euclid's fifth postulate is actually a proposition about parallel lines. It is much more complicated than any of Euclid's other postulates; and if Euclidean geometry is regarded as an abstraction of sensory experience, there is no apparent reason for believing that the fifth postulate should be universally true in that experience. It might be found, for instance, that in measuring very great distances, such as those in astronomy, experience would contradict some equivalent of the fifth postulate. One such equivalent is the useful proposition which Descartes saw as an eternal necessity: the sum of the angles of any plane triangle is equal to two right angles. Gauss actually suggested an astronomical test of this proposition as a means of deciding whether Euclid's geometry is an accurate account of the "space" known to experience. For reasons into which we need not enter, the test was never made, and if it had been, it could not have been sufficiently accurate to settle anything.

A still more plausible equivalent of Euclid's fifth postulate than the one mentioned was noted by Saccheri. It is one of three mutually exclusive propositions exhausting the possibilities with respect to parallel lines. Instead of stating Saccheri's equivalent, we may see the heart of the matter in a yet simpler and even more plausible equivalent of Euclid's postulate, and therefore also of Saccheri's.

A point p and a straight line l not passing through p fix exactly one plane in space. Imagine the sheaf of all the (straight) lines lying in this plane and passing through p. There are precisely three possibilities: exactly one line of the sheaf does not intersect *l*; more than one line of the sheaf does not intersect *l*; no line of the sheaf does not intersect *l*.

The first of these three is equivalent to Euclid's fifth postulate. It is therefore also equivalent to the proposition which Saccheri undertook to deduce from Euclid's other assumptions. He proceeded to convince himself that each of the second and third possibilities (or, rather, his respective equivalents of them) leads to a contradiction. From each in turn he had deduced a chain of consequences. So long as he trusted his own keen reason and kept his will to believe in subjection, he failed to reach the ardently desired contradiction in either of the non-Euclidean equivalents. His rigidly circumspect logic kept turning out nothing but consistency. This would never do.

Either deliberately or by an understandable oversight, the desperate believer in Euclid disposed of one of his new geometries by slipping in an additional postulate which he neglected to state: It is false that a straight line if sufficiently prolonged returns into itself and is of finite length. The other he succeeded in rejecting by an improper use of the infinitesimals. Ignoring the rules of the game he had undertaken to play straight, he relinquished the prize he might have had for the taking. It was already in his hand when he resigned. But as he was subconsciously determined to win for Euclid before the game began perhaps he could not help himself. One of the two non-Euclidean geometries he let slip through his grasp seems to have tempted him strongly. He dismissed it with evident regret. It was the one which Lobachewsky was to come upon ninety-seven years later.

Serenely confident that he had established the necessity and eternal truth of Euclid's geometry for all time, Saccheri entitled his masterpiece *Euclid cleared of all blemish* (*Euclides ab omni naevo vindicatus, etc.*). Almost from the time of Euclid himself ingenious geometers had striven to deduce Euclid's fifth postulate from his other assumptions, and all had failed. We know now that failure was inevitable: the fifth postulate is independent of the others, as was shown unwittingly by Saccheri, consciously by Lobachewsky, in their creation of non-Euclidean geometries. But Saccheri died happy in his disbelief in the real greatness of his own work.

If Saccheri's intellectual life was a tragedy it was at least not squalid. The same cannot be said for what some anticlerical commentators assert is the real tragedy of Saccheri's masterpiece. It was not lost or ignored for more than a century and a half. It was suppressed. This is an unpleasant insinuation, and our only purpose in discussing it at all is merely to enhance the historical importance for all "truth," from mathematics to theology, of the appearance of non-Euclidean geometries in the eighteenth and nineteenth centuries.

We saw that during the Renaissance, Euclid's geometry joined the eternal truths. Whoever might be rash enough to question the absolute necessity of this particular geometry could reasonably count on being housed with the heretics, if not less uncomfortably with the lunatics. Some absolutes, demonstrably necessary to any sane reasoning mind, there must be; Euclidean geometry was the chosen exemplar of all. Skeptics who suspected other works of the pure reason, in particular the official theology, might be discomfited by geometry. No man in his right mind could dispute the absolute truth of geometry. Therefore one absolute truth existed. And if one, why not two? But should some heretic overthrow the absolutism of Euclidean geometry, as Copernicus had overthrown that of Ptolemaic astronomy, no absolute would be secure.

As late as 1920, when relativity was sweeping the scientific world like a conflagration, upholders of the mediaeval tradition declared that if Euclid's was not the absolute and unique geometry of the universe, then the Holy Scriptures were in jeopardy and the road was wide open to atheism. The geometry of relativity was nothing like Euclid's. It had been familiar to mathematicians for about sixty years when Einstein found a scientific use for it. But the staunch adherents of the Middle Ages had never heard it mentioned. Nor were they aware of Saccheri's non-Euclidean geometries, then approaching their two-hundredth anniversary, which had been publicized for thirty years. The storm subsided, or its clamor was drowned in the uproar of the tempest descending on the absolutes of classical logic—a remote but direct consequence of the abolition of geometrical absolutism.

The facts concerning Saccheri's masterpiece are plain. On the recommendation of a Provincial of the Jesuits, the *Euclides*, having passed the scrutiny of a jury of theologians, was transmitted to the Senate, a Cardinal, and the Inquisitor General. The Inquisitor certified that the book contained nothing inimical to the official faith. Permission to print was granted on August 16, 1733. Saccheri died on October 25 of the same year. Any bookseller will explain the distinction between printing and publishing a book. Though the *Euclides* was printed in 1733, publication was delayed till 1889—the year in which Bruno's admirers succeeded in erecting their monument to him. From 1733 to 1888 the book lay buried. But in 1889 a copy was accidentally unearthed by Father Angelo Manganotti, of the Society of Jesus, who immediately recognized its historical significance. He brought it to the attention of a prominent secular geometer, Eugenio Beltrami. Brilliantly advertised by Beltrami, Saccheri's *Euclides* secured to its author all the rights and privileges of mathematical immortality a bare one hundred and fifty-six years after he was dead.

By 1889, however, non-Euclidean geometries, Saccheri's pair among them, were commonplace. They had been before the mathematical world for sixty years; and any man with some training and a little imagination could easily construct others differing from those already in existence. If the honor was of any importance to him in 1889, Saccheri could not claim scientific priority for his work of 1733, the accepted rule of priority in science being first publication. This may be unjust, but it prevents wasteful quarrels.

Another great geometer, W. K. Clifford (English, 1845-1879), called Lobachewsky the Copernicus of Geometry. Had Clifford known that Lobachewsky's non-Euclidean geometry of 1826-1829 was to come to light in the earlier work of an obscure Jesuit professor, whose name might have been in every history of mathematics but was in none, he could have called Saccheri the Copernicus of Geometry. Indeed the title in some ways is more appropriate for the Italian than the Russian. Copernicus received the first printed copy of his book, which overthrew the Ptolemaic system of astronomy, on his deathbed and so escaped official displeasure. Saccheri almost duplicated this feat. But whereas Copernicus' book was printed and published, Saccheri's was merely printed.

In a period of suppression of free thought such as that which began in the early 1930's in Germany, it may be of interest to state the reasons why some critics believe that the disappearance of Saccheri's printed book was no slovenly accident. If any moral is implicit in this admittedly conjectural history, it may be that suppression is not merely futile but stupid. Facts, like embezzlements, will out; and whoever tries to conceal them sooner or later is shown up as a blundering incompetent.

When some potential heretic of the Renaissance wished to circumvent the authorities he would pretend that his discoveries, scientific or other, were in no sense true but simply amusing fictions. Orthodox amateurs in authority would then —occasionally—let the alleged fictions pass as such, while unorthodox experts would see through the farce and study the subversive new doctrine diligently. Galileo resorted to this device, and might have been left unmolested had his love of satire not seduced his good sense. It is supposed that Saccheri tried a similar trick.

After seventy quarto pages of irreproachable reasoning, Saccheri casually tossed the more interesting of his new geometries aside with the inexplicable comment that it is obviously false. Either he was determined to sacrifice his reason to his faith in Euclid, or he dared not confess his belief in the heretical geometry. This sudden repudiation of sound logic jarred the unclerical Beltrami's sense of the fitness of things. A logician of Saccheri's acuteness, he imagined, simply could not have reached such a conclusion while his reason still functioned. Why then did he pretend that he had? The answer is immediate: fear. Saccheri dared not hint that the new geometry was "true." The flawless geometer Euclid was almost as sacred as that infallible logician Aristotle himself to Saccheri's superiors. To deny Euclid would be tantamount to questioning the classical logic by which the basic dogmas of the official theology had been established for all eternity. To have claimed that a non-Euclidean system was as "true" as Euclid's would have been a foolhardy invitation to repression and discipline. The Copernicus of Geometry therefore resorted to subterfuge. Taking a long chance, Saccheri denounced his own work, hoping by this pious betrayal to slip his heresy past the censors and get it into print. The trick—if such it was—worked. The book was printed.

If the *Euclides* was as false as Saccheri said it was, in what may have been a desperate hope that his epochal invention should not perish with him, it was, nevertheless, altogether too suggestive a book to leave lying about within reach of the young. So clear and convincing was the reasoning in Saccheri's new geometries that almost any rational mind might succumb to improper thoughts while following the alluring proofs. Whether or not the book was quietly suppressed, as a matter of conservative policy it should have been, in the interests of immediate security. Its teachings were wholly subversive to some of those of its sponsors; and if an organization is divided against itself, its chances of survival are slight. But in all such crises of expediency the suppressors leave time out of their opportunistic calculations. They overlook the possibility that some free mind beyond their jurisdiction will independently come upon the objectionable discovery and publish it to the world, thus robbing them of the honor they might have had for a little courage. It was so with Saccheri.

When the *Euclides* finally made its first public appearance in 1889, non-Euclidean geometries were established members of the mathematical hierarchy. No dreaded outburst of religious skepticism had followed their advent. Even the professional mathematicians were slow to recognize what the co-existence of several distinct, mutually incompatible, self-consistent geometries might mean for the future of the Platonic realism of mathematical truths in which nearly all of them still believed. A profounder revolution than the overthrow of the Ptolemaic astronomy had occurred almost unnoticed. All thinking, not merely an antiquated description of the solar system, was to be affected by the overthrow of Euclidean absolutism. What had been unthinkable before Saccheri constructed his geometries was to become the working creed of thousands whose business it is to think in order that others may act: the truths of mathematics and the mathematical formulations of the principles of science are of purely human origin; they are not eternal necessities but matters of human convenience. Neither in mathematics nor in science are there any absolutes.

From this beginning the disbelief in eternal truths and absolutes spread, but gradually, to logic and metaphysics, and from them to all authoritarianism. In a sense that Henley never intended, his boast "I am the master of my fate, I am the captain of my soul," at last began to have significance, and the "eternal spirit of the chainless mind" began to mean something. The human mind was as free as it willed to be, and the human race had now the opportunity to put away childish things and make of itself what it would.

Possibly those who mislaid Saccheri's *Euclides* foresaw what might happen to all absolutes should the work be published, and dreaded to precipitate a premature realization of the inevitable. Others had made a similar mistake regarding the Copernican revolution. Instead of repeating the tactical blunder, the careless Inquisitor responsible for the loss of Saccheri's work might have redeemed his predecessors by boldly proclaiming a revolution more subversive than the Copernican. He might even have honored Saccheri, his own subordinate, with the well-merited title of the Copernicus of Thought.

The life of the man who at last got a non-Euclidean geometry before the world is another story of success in things of relative unimportance crowned by personal frustration in a life's ambition. Though aware of the wide significance of what he had done, Lobachewsky died practically unrecognized by those capable of appreciating his work, and disgraced by the petty bureaucrats for whom he slaved.

There is no need here to list all the men who attempted to deduce Euclid's postulate of parallels from his other assumptions. The astronomer Ptolemy in the first century A.D. was one of the earliest, but even he had predecessors. He was followed in the ninth to the thirteenth centuries by several ingenious Moslem geometers who got no farther than he, among them the Persian mathematician and poet Omar Khayyám. He started along the very way that Saccheri was to choose. But he did not get very far. The Moslems in their turn were succeeded by yet more ingenious Italian geometers of the sixteenth and seventeenth centuries, who likewise arrived at no decisive conclusion. Several, including the notable English mathematician John Wallis (1616-1703) in 1693, replaced Euclid's fifth postulate by some equivalent assumption. Forty years after Wallis' attempt, Saccheri lost himself in the same blind alley that had swallowed up all of his predecessors, though he proceeded with incomparably greater circumspection than they. But he too believed that Euclid's assumption is necessary. In his search for truth, Saccheri, like all of those who had gone before, lacked the boldness or the imagination to execute an about-face and simply walk out of the alley that led nowhere. Merely to suspect that the required proof of Euclid's parallel postulate is impossible took as much daring and as much imagination as Copernicus exercised when he displaced our planet from the center of the solar system. Nor was the ability common to validate the suspicion by deliberately creating a geometry in which Euclid's postulate is rejected.

Lobachewsky (1793-1856) was endowed with the requisite ability. So also was his young contemporary, the Hungarian cavalry officer and geometer Johann Bolyai (1802-1860), who likewise had the necessary qualities of daring and imagination. Unknown to one another, Lobachewsky and Bolyai followed convergent roads to the same goal, reaching it almost simultaneously. The Russian has priority of publication. A few others about the same time almost succeeded in producing a consistent geometry other than Euclid's. But in addition to daring, imagination, and ability, a fourth prerequisite was essential to success: courage. Unless the man who produced, or claimed to have produced, a new and revolutionary geometry had the stamina to stand up for his work before wise men and fools, he might as well never have done it for any influence it might have. Fearing "the outcry of the Boeotians," Gauss, who had succeeded at least as early as the others, suppressed his researches. Lobachewsky and Bolyai did everything in their power to advertise what they had done. As Lobachewsky got his work into print first we shall attend to him alone, understanding that Bolyai is entitled to equal credit.

Lobachewsky came up the hard way. At the age of seven he lost his father, a minor official of the Russian government, who left his widow two sons and little else. The mother succeeded in putting her boys through school, and in 1807 the future mathematician entered the University of Kazan. For the next forty years of his life Lobachewsky was to be a member of the University, passing through the several academic grades from student to professor of mathematics and, finally, rector. Forty years of distinguished service to science and the promotion of higher education in Russia were rewarded by a purge. For no stated reason Lobachewsky was abruptly dismissed from his position at the age of fifty-four. Though his colleagues unanimously protested what they considered a bureaucratic outrage, the government stood by its action and refused to offer any explanation.

Lobachewsky lived another nine years, dying unrecognized as the constructive rebel he was, in 1856. The first notice of his non-Euclidean geometry had been presented to a scientific society at the University of Kazan in 1826. It was mislaid; but in 1829-30 a sufficient exposition was written out anew and published in Russian. A German translation followed in 1840. Neither edition made any noticeable impression on the mathematical public. The one mathematician in the world—Gauss—who could have secured Lobachewsky's geometry the attention it deserved, praised it highly in private correspondence but that was all. Undiscouraged, Lobachewsky continued developing his non-Euclidean system, calling it pangeometry. The year (1855) before his death, the University of Kazan celebrated its semicentennial anniversary. To do unmerited honor to the occasion Lobachewsky attended the ceremonies to present a copy of his *Pangeometry*, the summary of his scientific life. The work was written in French and in Russian, but not with his own hand, for he had gone blind. A few months later he died at the age of sixty-two, possibly the only man in the world at the time who fully realized the significance of what he had done. Lobachewsky saw beyond the new geometry to its repercussion on all deductive reasoning. The last is the item of importance here.

The source of Lobachewsky's total success was his ability to disbelieve something that seemed to be necessarily true and his capacity for implementing his disbelief. This flair for constructively doubting the traditionally obvious seems to be the rarest of all intellectual gifts. Whoever has it, and exercises his talent, usually achieves a revolution.

When asked how he came to invent relativity, Einstein answered, "By challenging an axiom." Lobachewsky challenged Euclid's axiom of parallels; Copernicus challenged the axiom that the Earth is at the center of the solar system; Galileo challenged the axiom that the heavier of two bodies falls the faster; Einstein challenged the axiom that events at different places can be simultaneous; Brouwer challenged the axiom that Aristotle's logical law of the excluded middle is universally applicable; the atomic physicists of the twentieth century challenged more than one axiom of Newtonian mechanics; and there have been—and are—others. In each instance some department of human knowledge was transformed, and almost invariably in the direction of greater freedom. An axiom generally implies a compulsion of rational thought or a restriction of possible action; abolition of the axiom as a necessity invites free invention. In the past, abolition of axioms also invited persecution; today in all sciences except the social it merely invites personal abuse, if even that. Success in the acquisition of new knowledge or novel conveniences, consequent on the successful challenging of some venerated axiom, is usually a passport to respectability, till the newly constituted freedom itself becomes a tyranny, is challenged, and gives way to another. But the net residue is on the side of human freedom, not on that of inherited absolutes and vested traditions.

Lobachewsky's success in challenging an axiom was followed by others. It cannot be claimed that his example inspired the earlier successes following his, for his work lay almost neglected for nearly thirty years. In the 1840's, for example, William Rowan Hamilton challenged one of the basic axioms of classical algebra. It had been supposed that the axiom "the order in which two quantities are multiplied together does not affect the result" is necessary for a consistent algebra. In the algebra developed by Hamilton for applications to the physical sciences this axiom was rejected. It seems strange that Hamilton lived for nearly thirty years after Lobachewsky published his work and yet died ignorant of the existence of non-Euclidean geometries. But in the light of Hamilton's personal success this only shows that the creative mathematicians as a class were at last beginning to recognize the inherent freedom of their efforts. When at length the deeper significance of the work of Lobachewsky and Hamilton was appreciated, the challenging of axioms in mathematics became one of the commoner methods of making advances. Free invention flourished unrestricted by tradition, and mathematics entered a period of unprecedented expansion. Toward the close of the nineteenth century, Georg Cantor could express the conviction of a majority of creative mathematicians in an aphorism now famous: "The essence of mathematics is in its freedom."

Agreed, say the Platonic realists. But what is this freedom of mathematics? Were not all the queer geometries and freakish algebras already in eternal existence before mathematicians "discovered" or "observed" them? Were they not unknown to mortals simply because mathematicians were blind to reality? Against such an obstinate will to believe in the indemonstrable and the forever unattainable, not to say the useless, rational skepticism is impotent and common sense must strive in vain. Let who will believe, say the naturalists.

Those who persist in their belief that "mathematical reality lies outside us" have at least one unanswerable argument on their side. Invention may be free, they admit, but free only within the law. That law is logic as it has developed in mathematics since the time of Thales. But, as we have seen, this supposedly rigid law is itself subject to change. This is no disaster for the resolute realist: the change itself is subject to a higher law, which in its turn is subject to a higher, and so on, to that unattainable highest which is the Absolute.

The freedom which Lobachewsky imagined he enjoyed in creating his geometry was an illusion: the Absolute dictated the geometer's every move. The "essence of mathematics" is not, as Cantor declared, in its freedom, but in its slavery to a despotism forever beyond any assault by human beings. Again let him believe who will.

CHAPTER 26

Changing Views

FEW philosophers have succeeded in resisting the temptation to apply their philosophies, sometimes with disastrous results, to the rudiments of mathematics. The simplest arithmetic and the beginnings of geometry have seemed to the majority of metaphysicians to be necessities for any coherent account of the physical universe. So unless some professedly comprehensive system of knowledge could explain this apparent inevitability of numbers and elementary geometrical theorems, it remained unconvincing.

A reasonably ambitious metaphysics had also somehow to rationalize those persistent mysteries "space" and "time." Otherwise the physical sciences would be left without a philosophical foundation. If in addition "space" and "time" could be linked with the accepted geometry, arithmetic, and physics of a particular epoch, the corresponding metaphysics would have comprehended nearly everything. When the "laws of thought"—Aristotle's classic triad of identity, excluded middle, and contradiction—were also included in the supreme synthesis, the metaphysics for the entire philosophy, with the exception of two capital details, was complete. A theory of ethics, morals, and values, and an argument for the existence of the deity had to be provided within the system.

All this Immanuel Kant (1724-1804) achieved. If certain parts of his colossal success are less impressive to the mathematicians and scientists of today than they may have been to their predecessors of the eighteenth and nineteenth centuries, it is because both science and mathematics are now more dynamic than they seemed to be in 1781 when Kant published his *Critique of Pure Reason*. Whatever the date, neither the science nor the mathematics of today is wholly that of yesterday. Kant himself no doubt recognized half of this truism as a decisive factor in the obsolescence of universal philosophies, when he declared that nothing had been more injurious to philosophy than mathematics. As Kant's was for long the most durable of mathematical philosophies since Plato's, lasting well into the nineteenth century, we shall describe it in some detail as our parting tribute of respect to the great philosophers of the past.

Before considering Kant's most ambitious contribution to the critiques of science and mathematics, we may glance at the personality and career of this foremost of modern philosophers. Except for the possible defect that he was clean-shaven all the sixty years he might have grown a beard, Kant is the flawless museum specimen of the popular ideal of a professional philosopher. He has been called a pedant, but hardly with justice; pedants are never even abortively creative. All he had went into thinking. To Descartes' "I think, therefore I am," Kant might have retorted, "I am, therefore I think."

Blessed, like Descartes, with frail health in his youth, Kant learned at an early age to respect his body, and when he grew up took such extremely good care of himself that he lived to the age of eighty. Either for lack of initiative and inclination or because he grudged the time, he had no traffic with women after he escaped from his excessively pious and somewhat domineering mother. What, if any, his relaxations were, beyond thinking about thinking, is unknown. In his later years he indulged in oracular utterances and was not averse to being revered in public. He enjoyed the rare distinction of becoming a legend while still in the flesh. If unswerving devotion to a single purpose is to be admired in a human being, Kant was one of the most admirable mortals that ever lived. His goal was the creation of a philosophy that would outlast him, and this he attained. It seems to be generally agreed that his success was as arid as his personality. Though he philosophized much on aesthetics he had no artistic or literary sense whatever. As for the details of his material life, they might fit almost any overworked, underpaid professor of philosophy in any second-rate college or university today. All that this greatest of philosophers since Plato had was his greatness.

The metaphysically and theologically inclined Scotch claim Kant as one of their own. Kant's father spelled his name Cant. The German Kants are said by some partisans to be descended from the Scotch Cants who had migrated from Scotland to Prussia. It is claimed that Kant's paternal grandparents were Scotch. Probably it is no longer of any great consequence whether or not this claim is justified. However, in the heat of the First World War, when the combatant nations were trying to prove that none of their enemies had ever produced a man worth remembering, Beethoven was assigned to the Flemish or the Dutch, Kant to the Scotch, and Gauss to the Jews. The Second World War being more nearly total than the first, such diversions were neglected for less academic disputes.

Whatever his origin, Immanuel Kant was born in 1724 at Königsberg, in the Kingdom of Prussia, one of eleven children of a hard-working saddler rich in piety but poor in goods. Of the four sons only Immanuel and a brother eleven years his senior survived beyond infancy. Immanuel's parents, especially his mother, made grinding sacrifices to have their more studious boy trained for the Lutheran ministry. At the age of seventeen Immanuel entered the University of Königsberg determined to repay his parents by mastering all theology. The liberality of the academic atmosphere—contrasted with that of his super-pious home—caused him quickly to change his mind. For theology he substituted mathematics and philosophy, an ominous combination. His student career was phenomenal. To an unprecedented acquisitiveness he added an inhuman diligence. By attending innumerable lectures and incessant reading in science, theology, and the classics of antiquity, the insatiable young man transformed himself into an encyclopaedia of the erudition of his time.

Among other items to which Kant applied his persistence was the *Principia* of Newton. From the future philosopher's personal adventures in science it seems doubtful that he could have got much out of his reading of Newton, and some critics believe the Newtonian natural philosophy had a deleterious effect on Kant's metaphysics. One thing however is certain: Kant's philosophy did not retard the progress of Newton's mathematical astronomy.

As a result, possibly, of his poring over the *Principia*, Kant imagined (1755) the famous nebular hypothesis to account for the origin of the solar system. This daring conjecture was the high point of the philosopher's scientific career, and he took it with corresponding seriousness. Anything approaching an adequate test of its reasonableness was far beyond the mathematics of Kant's day and even, apparently, of our own. Laplace, the foremost mathematical astronomer of the eighteenth century, also suggested the nebular hypothesis, but with a most significant difference. He realized that there was no scientific foundation for his picturesque guess and tossed it off almost as a joke. This episode of Kant's foray into science has been recalled to illustrate the fact that speculation unbacked by recognized scientific procedures is less difficult than the painstaking framing and rigorous testing of hypotheses as practiced by professional scientists. Though an untestable speculation born of intuition may enjoy all but immortal popularity, it is not on that account truer or more valuable than some scientific hypothesis proposed on a Monday and abandoned the following Wednesday.

Kant's academic course was singularly devious for a man of his great and recognized genius. After his father's death the young man's monetary circumstances were such that he was forced into private tutoring for nine years. Through the generosity of friends who perceived his destiny and believed in him, he was enabled at last to return to the University, where he took his doctorate at the age of thirty-one. Then followed a probation of fifteen years at the University of Königsberg as a Privatdozent—an unpaid instructor who lived by hard labor and what fees he could collect from voluntary students. Kant offered crams in practically all subjects in the curriculum to any who could afford to pay. He also wrote extensively on much of what he taught. All this thankless drudgery no doubt was of some benefit to the slowly ripening philosopher. It at least gave him an outlook over all human knowledge as it was in his time. The breadth and discursiveness of Kant's training for philosophy recalls Plato's happier effort to educate himself for that comprehensive and elusive profession.

The authorities recognized Kant's talents and tried to make it easier for him to exist. They even went to the extreme of offering him (1762) the professorship of poetry. Knowing himself, Kant had the good sense to decline. The following year he accepted the more suitable position of assistant librarian. At last, in 1770, when he was forty-six, Kant was appointed to the coveted professorship of logic and metaphysics he should have obtained twenty years earlier. The next eleven years were swallowed up in the laborious composition of his masterpiece, the *Critique of Pure Reason*, published in 1781.

Eleven years later Kant permitted himself the one serious fight of his otherwise peaceful life. The philosopher emerged morally victorious but humanly disgusted. His only gain was that he had proved to himself and to his friends that he, the puny and sickly professor, had more backbone than all the sturdy bigots in Germany, including the King of Prussia.

Though heretics were no longer being converted in the torture chamber or redeemed at the stake, Kant took as great a risk as Galileo or Bruno. Some administrative genius had discovered that innovators may be eliminated without awkward publicity by separating them from their food supply. In Kant's case it would have been sufficient to oust him from his professorship. Had he been expelled from the University as a dangerous radical or blasphemous atheist, his academic career would have ended then and there. He was competent only for university life or private teaching; and it is highly improbable that any patron liberal enough to entrust his son's education to a disgraced professor could have been found immediately in the Prussia of the 1790's. Knowing well what he was about and what the outcome might be, Kant fought for the freedom of his mind.

The main battle lasted nearly five years. The *Critique* had precipitated a holy war of more than usual ferocity. The unworldly metaphysician, who as a young student abandoned theology to learn something of mathematics and philosophy, had been sanguine enough to imagine that he might construct "religion within the limits of reason," and at the same time delight all the orthodox Lutherans in Germany. Kant's error was the strictly rationalized eighteenth-century variant of the delusion that misled Augustine and his successors in their attempt to subjugate the deity to numerology. To his consternation Kant discovered that he had unchained the devil.

It seems to be generally admitted today that a religion wholly within the limits of reason is not as desirable as the unemotional Kant supposed it might be. He fought valiantly for his bloodless creed, but his opponents, including the morbidly pious King, were far too numerous and too well organized for any lone champion of reason to beat off. Yet without his own consent, grudgingly given to avert a more disastrous controversy, Kant could not have been silenced. The King's death five years after Kant had pledged himself to stir up no more animosity than was necessary to maintain his intellectual integrity, released the philosopher's tongue to say what it must. But five years of threats and repression had done something to Kant, and the fight had staled. What he had learned of the orthodox mind as it was in the Prussia of his day seems to have discouraged him from further attempts to enlighten it. He went on with the work for which he was created, addressing his further critiques to the few patient enough to unravel the intricacies of his thought.

Kant's attempt to settle once for all the status of mathematical truth is the only detail of his system of interest to us here. But it may be recalled that mathematics for Kant was almost as important as it had been for Plato. So if he was mistaken in his estimate of mathematics, it is possible that he was not entirely right in all other details of his vast system.

Kant's opinions on the nature of mathematics are expounded in *The Elements of Transcendentalism*, opening the second part of the *Critique*, and most explicitly in the *Transcendental Aesthetic*. He seems to have had some doubts whether he had made himself as clear as he and the "resolute reader" to whom he dedicated his conclusions could have wished. To placate his conscience he composed an explanatory sequel, intended particularly for any teachers who might be hardy enough to adopt the *Critique* as a textbook. The sequel is titled, modestly enough, *Prolegomena to Every Future System of Metaphysics which can claim rank as a Science*. There were giants in the Age of Reason. Among the "General Questions" raised and supposedly answered in the *Prolegomena* two are relevant here: "Is metaphysics possible at all?" and "How is pure mathematics possible?" Kant's answer to the first, as might be anticipated, is Yes. The extreme logical positivists of the twentieth century assert that the correct answer is No.

Kant's question about pure mathematics is the one of immediate interest. His comprehensive misunderstanding of the nature of mathematics is sufficiently exemplified in his false proposition, by which he set great store, that geometry consists of "synthetic judgments *a priori*." It will be enough to describe what he meant by this and to point out why mathematicians know—it is a matter of knowledge, not of opinion—that it is erroneous. To anticipate, Kant was misled by the distinction, not clearly recognized when he composed the *Critique* but now commonplace, between geometry as an abstract deductive system and geometry as a partly empirical science applicable to the physical universe. He was similarly deluded about arithmetic, and indeed about all mathematics. As Einstein expressed the difference between applied and pure mathematics: "So far as the theorems of mathematics are about reality they are not certain; so far as they are certain they are not about reality."

It is not to Kant's discredit that he overlooked this fundamental distinction. With the exception of Saccheri's buried non-Euclidean geometry, of which Kant was unaware though it had been in print for forty-eight years when the *Critique* was published, the mathematicians had scarcely provided the philosophers with material enough on which to base an intelligent opinion. And we saw how slow the mathematicians themselves were to appreciate the significance of Lobachewsky's non-Euclidean geometry, published a quarter of a century after the death of Kant. It was only toward the close of the nineteenth century that professional mathematicians became seriously interested in the nature of mathematics, and began then to understand the nature of what their predecessors from Thales to Poincaré (1854-1912) had accomplished.

It is only fair to hear what Kant himself said before we pass to denials. The few excerpts that follow will suffice. He begins with a definition: "I call all representations in which there is nothing belonging to sensation *pure*. The pure form therefore of all sensory intuitions, that form in which the several elements of phenomena are seen in a certain order, must be found in the mind *a priori*. And this pure form of sensibility may be called the pure intuition." After some further definitions Kant states that "In the course of this investigation it will appear that there are two pure forms of sensory intuition as principles of *a priori* knowledge, *Space* and *Time*. What then," he asks, "are Space and Time? Are they real things? Or, if not, are they determinations or relations of things, but such as would belong to them even if they were not perceived? Or are they determinations and relations inherent solely in the form of intuition, and therefore in the subjective nature of our mind, without which such predicates as space and time would never be ascribed to anything?"

Before hearing Kant's answers to these questions we may transcribe two explanations from the dictionary. "Kant . . . held that *a priori* knowledge consists of certain 'presuppositions' (as space and time) and principles of understanding that are antecedently necessary in order that experience in general should be intelligible." This fixes the constantly recurring *a priori* to which Kant appeals. The other technical word is "apodictic," which means "involving or expressing necessary truth; absolutely certain; also capable of clear and convincing demonstration." Granting that these pregnant definitions are clear—though hardly so perspicuous as those of the elementary geometry to which Kant applied them—we may perhaps understand what was intended by the following summary. Kant stated his conclusions in four general propositions, of which only the capital parts need be given.

"1. Space is not an empirical concept which has been derived from experience. . . . The representation of space cannot be borrowed through experience from relations of external phenomena, but on the contrary external phenomena become possible only through the representation of space.

"2. Space is a necessary representation *a priori*, forming the very foundation of all external intuitions. . . . Space is therefore . . . a condition for the possibility of phenomena, not . . . a determination produced by them. It is a representation *a priori* which necessarily precedes all external phenomena."

Because the next is so comprehensively and elaborately wrong in the light of current knowledge it is transcribed in full.

"3. On this necessity of an *a priori* representation of space rests the apodictic certainty of all geometrical principles, and the possibility of their construction *a priori*. For if the intuition of space were a concept gained *a posteriori* [only from experience], borrowed from general external experience, the first principles of mathematical definition would be nothing but perceptions. They would be exposed to all the hazards of perception, and there being (for example) only one straight line between two points would not be a necessity, but only something taught in each instance by experience. Whatever is derived from experience possesses only a relative generality, based on induction. We should therefore be unable to say more than that, so far as hitherto observed, no space has yet been found having more than three dimensions."

We shall recur to some of this presently. Kant's fourth and last general conclusion adds little to the first three.

"4. Space is . . . a pure intuition. . . . An intuition *a*

priori, which is not empirical, must form the foundation of all conceptions of space. In the same way all geometrical principles, for example 'any two sides of a triangle are together greater than the third,' are never to be derived from the general concepts of side and triangle, but form an intuition, and that *a priori*, with apodictic certainty."

This (Kantian) account of space and geometry, with its *a priori* intuitions and its apodictic truths, was long favored as final by numerous metaphysicians. Kant formulated a similar doctrine of numbers and arithmetic. Both have been abandoned because neither accords with mathematical fact. His *a priori* "time" has gone the way of his geometry and arithmetic, not because it disagrees with mathematics, which is not concerned with speculations on the nature of time, but because it is contradicted by modern experimental and theoretical physics. We need indicate here only what disposed of Kant's geometry.

Kant believed that geometry consists of statements ("*judgments*," declarative sentences) which are independent of experience (are "*a priori*") necessarily true ("*apodictic*"), and which contain factual content (are "*synthetic*"). That no such statements exist in mathematics (or elsewhere, so far as is humanly known) is one of the simpler conclusions of modern mathematical logic. Kant's confusion originated in his failure to distinguish two radically different things. His readers' bewilderment may arise from his ineffectual struggle to expound both things simultaneously and in the same words without being aware that he was talking, not of one thing, but of two. His pair were "physical geometry" and "mathematical geometry."

Physical geometry in its practical aspect is a partly empirical science designed to give a coherent account of the world of sensory (and scientific) experience. Mathematical geometry is a system of postulates and deductions from them, designed without reference to, or intended relevance for, sensory experience. In mathematical geometry "truth" is identified by one school of modern mathematical philosophers with consistency (freedom from contradiction within the system); in physical geometry "truth" includes also approximate agreement with observable phenomena. When sufficiently analyzed the statements of mathematical geometry are "true" merely by their form as logical sentences. Such sentences are called "analytic"; for example, "It is raining now or it is not raining now." But "It is raining now" is a definite one of "factually true," "factually false," and which one it may be is ascertainable by going outside to look. This sentence has factual content. The first has not; it says nothing about the actual weather.

The distinction between physical and mathematical geometry may be illustrated by Kant's unfortunate examples in his third general conclusion quoted above. If "straight line" is intelligibly defined, it is not so in either mathematical or physical geometry that there necessarily is only one straight line between two points. Euclid's definition is "A straight line is a line which lies evenly between its extreme points"; and Kant may have been thinking of this vaguely intuitive notion. A moment's reflection will show that Euclid's supposed definition fails to define anything at all. As frequently stated in school geometries, "A straight line is the shortest distance between two points." This is both intuitively satisfactory and useful when, as at a somewhat later stage, "point" and "distance" are given precise numerical definitions. To avoid a mystery where none is concealed, "straight line" is replaced by the word "geodesic." A geodesic in a "space" is either a least or a greatest distance between two points in that space. (This definition is close enough to the precise mathematical definition for our present purpose.) If the space considered is the surface of a sphere (not what the surface encloses, but the surface itself), diametrically opposite points may be joined by an infinity of geodesics (arcs of great circles on the sphere). "The hazards of perception" to which Kant alludes seem to have included, for him, the illusion that the Earth is flat. His second example, "no space has yet been found having more than three dimensions" was nullified long ago by the construction of spaces having any given number of dimensions. The scientifically useful space of four dimensions in relativity is perhaps the most familiar instance of a space having more than three dimensions.

It was unfortunate for the import of all his metaphysics that Kant singled out Euclidean geometry as an instance of a necessary truth. The distinction between mathematical and physical geometry followed the invention of non-Euclidean geometries. Each of these geometries, Euclid's also when obvious blemishes are removed, is self-consistent, and none is compatible with another. Each is mathematically "true." Which is physically "true"? It has turned out that more than one geometry is reasonably sufficient for certain scientific purposes, but that some particular one is more satisfactory than its competitors for other purposes. Each is "true," namely self-consistent, in the abstract, logical, or mathematical sense; any one of several is "true" in the physical sense for its appropriate range of phenomena; but no two, being incompatible, are factually true for the same range. When all this was first clearly recognized in the early 1900's, some scientists and mathematicians identified "truth" with convenience. But there was no necessity to introduce a new confusion into a situation from which confusions had at last been expelled after about two thousand years of misunderstanding.

It would be interesting to explore Kant's numerology, especially in his famous table of twelve categories—displayed as four trinities, one at each of the cardinal points. But we shall not take space to exhibit or to discuss his most suggestive trichotomies in which, for the first time, a philosopher departed from the Pythagorean "division by two"—dichotomy—and boldly divided by three. Instead we pass on to what Kant's leading mathematical contemporary, Gauss, already mentioned as one of the three greatest mathematicians in history, observed regarding the mathematical philosophy of Kant and other mathematical amateurs. First a word about Gauss himself.

Gauss (1777-1855) was twenty-seven when Kant died, and was already acknowledged by his nearest rivals to be the foremost mathematician in the world. If Kant ever heard of Gauss he made no mention of the fact. Both the mathematician and the philosopher were notorious stay-at-homes. Kant's longest trip from Königsberg was forty miles; Gauss' record was twenty-seven miles from Göttingen, and for each man the maximum was a unique adventure. In other respects "the greatest philosopher since Plato" and "the greatest mathematician since Newton" were strikingly different. The robust Gauss, blessed with sound health all his life, was a confirmed hypochondriac. The diminutive and frail Kant kept himself going only by the strictest self-discipline, and never seemed unduly concerned about his health. But they were alike in their irrational horror of death. When either lost a friend the departed one was dead indeed, mere mention of his name being prohibited. Gauss never courted veneration, and although fully aware of his own tremendous part in the development of modern mathematics, he had no trace of self-esteem. Kant in his later years made somewhat of a nuisance of himself with his pontifical infallibility. The metaphysician's mental faculties deteriorated with age; the mathematician's remained whole and vigorous to his dying hour.

The most radical difference between Kant and Gauss is the point of interest here. If a comparison between incommensurables like metaphysics and mathematics is possible, it is probable that Gauss understood more about metaphysics than Kant did about mathematics. After his student days Kant had no inkling of what was taking place in living mathematics. Gauss on the other hand was a close student of philosophy all his life. His command of languages enabled him to keep abreast of what was happening in philosophy not only in Germany but elsewhere. Of course he never claimed to be anything more impressive in philosophy than an interested amateur. But an amateur of the magnitude of a Gauss may conceivably be more worthy of serious consideration than almost any three professionals, specifically in the philosophy of mathematics. Here perhaps Gauss had an unfair advantage. Had he been ambitious for fame, he, rather than Lobachewsky, might have claimed the title of the "Copernicus of Geometry." His interest in the foundations of geometry had begun when he was a boy of twelve. When Kant died Gauss had already made some progress toward the non-Euclidean geometry which he deliberately suppressed to avoid profitless battles of words with mathematical bigots and metaphysical ignoramuses. So it is not remarkable that Gauss was unsympathetic to Kant's philosophy of mathematics. For in spite of his reading of Newton, or perhaps because of it, Kant's mathematical equipment for attacking the problem of mathematical truth was as antiquated as Euclid's. So far as its relevance for mathematics is concerned, the *Critique* might have been composed in the fourth century B.C., instead of in the eighteenth A.D.

Kant might have learned something from Berkeley's *Analyst,* had he not harbored sentiments suspiciously like professional jealousy for his idealistic rival. He could have learned nothing from Gauss, because "the prince of mathematicians" was never greatly interested in teaching anybody anything. Gauss hated all forms of instruction; his masterpieces were designed with economical completeness in mind, not with the reader's comfort. Comparatively few dipped into them, and fewer still got to the bottom of what they struggled to fathom. His tone in his published work was always one of strict justice or frigid politeness to his predecessors or contemporaries. But in his letters to trusted friends he could be as blunt as a peasant. A rather mild specimen from the year 1844 tells us what Gauss really thought of mathematical amateurs when they undertook to explain mathematics, and incidentally gives his opinion of one of the cardinal ideas of all Kantian metaphysics:

"You see the same sort of thing [mathematical incompetence] in the contemporary philosophers—Schelling, Hegel, Nees von Essenbeck—and their followers. Don't they make your hair stand on end with their definitions? Read in the history of ancient philosophy what the big men of that day —Plato and others (I except Aristotle)—gave in the way of explanations. And even with Kant himself it is often not much better. In my opinion his distinction between analytic and synthetic judgments is one of those things that either peter out in a triviality or are false."

So much for professional knowledge against amateur opinion. But both knowledge and opinion are subject to change.

Not all of Kant's ideas about the nature of mathematics obstructed progress, as progress was understood in the early twentieth century. We alluded in an earlier chapter to the intuitionist variety of mathematical philosophy initiated by Brouwer about 1912 and since extensively developed by Weyl and others. It was noted in connection with the mathematical infinite that Brouwer, following Kronecker (1823-1891), refused to admit that a proposition is either true or false unless some means for deciding which is prescribed. The intuitionists deny Aristotle's law of excluded middle where no such means exists. Incidentally this denial sets aside many of the long-accepted "existence proofs" of classical mathematics, both pure and applied. This however is not the detail of immediate interest. It is the philosophy accompanying the intuitionist creed which takes us back to Kant and his insistence on "intuition" in mathematics. Brouwer at first imagined that Kant's philosophy had suggested his own, but later emphatically denied any indebtedness and repudiated Kant and nearly all his mathematical metaphysics. As only the creator of modern intuitionism, if anybody, can say what inspired him, it is futile to argue the matter.

Mathematics for the intuitionists is identified (intuitively?) with "the exact part of our thought," and is antecedent to both logic and philosophy. The source of mathematics is asserted to be "an intuition that presents mathematical concepts and inferences to us as immediately clear." It is denied that this intuition is in any sense mystical; it is merely "the ability to treat separately certain concepts and inferences which appear regularly in common thinking." It will be interesting to confront this denial and affirmation with a little of what the dictionary states under "mysticism": "the doctrine or belief that direct knowledge of God, of spiritual truth, of ultimate reality, etc., is attainable through immediate intuition, insight, or illumination, and in a way differing from ordinary sense perception or ratiocination." The "objects" with which intuitionistic mathematics is concerned are said to be immediately apprehended in thought. Reminiscent of Kant's "synthetic, *a priori*" geometry though not quite the same, these objects of the intuitionists are independent of experience and have no existence independent of thought.

Ancient and mediaeval dogmas reappear also in the intuitionists' insistence on the human capacity for imagining a sequence of "distinct, individual objects" obtainable by adjoining objects indefinitely, one at a time, to those already imagined. Starting with "one" and the conceptual operation of "adding one," the intuitionist thus intuits the unending sequence of natural numbers 1, 2, 3, . . . by indefinite repetition of the operation. This capacity for intuiting an infinite sequence of numbers, as we saw, is anything but universal among primitives, or even among civilized peoples—unless, of course, this rather ad hoc "capacity" is by hypothesis latent though unobservable. A thoroughgoing "finitist" (like Kronecker) might assert that the intuitionists' infinite sequence has only a meaningless verbal existence on paper, not an intuitive existence in the human mind. The finitist rejects the infinite as a delusion inherited from outmoded philosophies and confused theologies—he can get on without it.

While he was still overawed by Kant, Brouwer adhered to Kant's creed that space and time are what Kant said they are. He later (1912) gave up *a priori* space, but clung more tenaciously than ever to *a priori* time.

The date, 1912, has been recalled to emphasize the fact that mathematical intuitionism is older by about fourteen years than the modern quantum theory of physics. That theory and the subsequent development of atomic physics caused theoretical physicists at least to refine their intuitive notions of "space," "time," "number," and "identity" as concepts adequate for the description of all observable physical phenomena, especially those for which the nuclei of the atoms are believed to be responsible. The philosophical situation is in some respect analogous to that which followed the appearance of non-Euclidean geometries in the description of nature after the successes of general relativity. Just as the classical geometry (Euclid's) was found inadequate for certain scientific purposes, so the traditional "time" and the rest have needed revision to keep up with an ever-advancing science. It would seem that if mathematics is not to revert to a sterile formalism it must pay some attention to the sciences in which it is used and continually revitalized.

Though the nucleus of an atom is rather cramped quarters for a meeting between some of the greatest mathematical philosophers in history, at least three of them appear to have squeezed into it somehow. In the process of accommodating themselves to a modern scientific environment, Pythagoras, Aristotle, and Kant have been forced to discard some of their most treasured convictions. Pythagoras has given up his universal "number," Aristotle his "identity," and Kant his "time." In macroscopic—large scale—phenomena there is no difficulty about identifying and counting the objects contemplated. For example, each pebble in a heap can be identified and distinguished from all the others, and all the pebbles can be numbered off one, two, three . . . , till the heap is completely counted. This illustrates one of the four imaginable possibilities concerning identification and counting. What of the remaining three? Physics has encountered instances of two. Only the conclusions need be stated here: the quanta of light are unidentifiable and uncountable; electrons are unidentifiable but countable. Thus in physics, at least, the immemorial label of "number" is no longer universally applicable. As for "space" and "time" they too lose their traditional universality when pursued to the atomic nucleus surviving, if at all, only as secondary and artificial "constructions" imposed for mere economy of language on mathematical descriptions of observable phenomena. Another step or two in the same direction and it may be found that these supposed necessities of thought are not even conveniences but outmoded encumbrances impeding the understanding.

To complete the record of the major changes induced in basic thinking by attempts to understand the nature of mathematics, we report the verdict of modern mathematical logic on the traditional philosophy. Writing in 1933, Rudolf Carnap, then associated with the famed "Vienna Circle," summed up the situation as follows:

"The majority of philosophers have bestowed but scant attention on [the new logic created by mathematicians since 1854 and especially since about 1890]. The distrustful reserve with which they have approached this new logic is rather surprising. The mathematical dress in which it is clothed may indeed be somewhat intimidating; but it stirs up a deeper hostility, which we are beginning to discern clearly; the distrust is born of the danger which threatens the position of the old philosophy. And indeed every philosophy, in the old sense of the word, whether it called upon Plato, or St. Thomas [Aquinas], or Kant, or Schelling, or Hegel, or whether it erects a new 'metaphysics of Being,' or a 'dialectical philosophy; appears before the inexorable criticism of the new logic as a doctrine, not false in its content, but as logically untenable, and therefore devoid of meaning."

This is plain enough. As we closed our account of the great mathematical philosophers with Kant, we were obliged to forego the pleasure of exploring Hegel's ingenious ideas on science and mathematics. To compensate for this omission we again quote Carnap: "Since all the 'laws' of logic are tautologous and empty [of factual content], they can tell us nothing at all about the real world. Any dialectical metaphysics—as Hegel's largely is—is therefore denied all legitimacy."

Naturally not all philosophers concur in this death sentence. But few, philosophers or others, will challenge the earlier prophecy (1925) of the Oxford philosopher C. E. M. Joad:

"If Mr. [Bertrand] Russell is right, most philosophy is meaningless; if he is wrong, we may still hope by the methods which philosophy has traditionally pursued to arrive at truth about the Universe. Whether he is right or wrong, however, it is certain that men will continue to philosophize, if only because of the ennobling and widening effect upon the intellect of philosophical speculation, and the deep-seated character of the instinct of curiosity to which it appeals." The only dissent comes from those scientists, historians of culture, and observers of human nature who question "the ennobling and widening effect upon the intellect of philosophical speculation." But here, as so often in our journey from the past to the present, the disagreement may be rooted in opinion rather than in knowledge.

Some will see in this ceaselessly changing flux of theories and beliefs a disheartening picture of instability and futile strife. These may be assured that the one permanent state of rest is death. Others will note that each change replaces some old knowledge by new, that some of what once seemed sound is no longer so, and that what is to come will probably differ from what now is. Except for the guess as to the future, all this is historical fact; and whether we bless it as progress or damn it as retrogression, we cannot talk it out of history. Those who can accept change may find contentment; those who cannot may be miserable. Surely there is nothing so tragic in change as those make out who would forbid it, if they could; it is at least not stagnation. If some eternal truth of the past is now known to be neither true nor eternal, we are the better off by the loss of two errors. And if neither science nor mathematics can show a finality anywhere, the like is true of philosophy. Yet none of the three has been condemned as valueless. Whoever desires permanence in anything must seek it elsewhere than in these—unless he happens to be a numerologist or a mathematical realist. Others will follow the changing patterns in the kaleidoscope of time as each slight turn of events rearranges the colors in new and unpredicted designs, some more intricate or more beautiful than any thus far remembered, others less so.

What of the mathematical realist with whom we set out on our exploration of the past? For him there is no significant change. Everything real now is essentially as it was since time began. Somewhere, somehow, the numbers and the truths of mathematics exist as they always were and always shall be. This is not a malicious misrepresentation of the realist's creed. Hundreds of confessions of faith as confident as this might be cited from the first four decades of the twentieth century alone.

As only men have had their say thus far, we shall let a woman speak the last word, and we shall quote excerpts from a thoughtful address (1925) by the distinguished English geometer Hilda Phoebe Hudson.

"To all of us who hold the Christian belief that God is truth, anything that is true is a fact about God, and mathematics is a branch of theology. . . .

"An old Greek, a French child, and a self-taught Indian, each finds for himself the same theory of geometrical conics. The simplest and therefore the most scientific way of describing this, is that they have discovered, not created, a geometry that exists by itself eternally, the same for all, the same for teacher as for taught, the same for man as for God. The truth that is the same for man as for God is pure mathematics as distinguished from applied . . .

"But however we think of heaven, it is hard to imagine astronomy and botany surviving as they are, and having much interest or importance there. . . . On the other hand it is just as hard to imagine pure mathematics not surviving.

The laws of thought, and especially of number, must hold good in heaven, whether it is a place or a state of mind; for they are independent of any particular sphere of existence, essential to Being itself, to God's being as well as ours, laws of His mind before we learned them. The multiplication table will hold good in heaven. . . .

"God shines through his works as clearly in logic as in matter.

"The whole of geometry is so filled with the glory of God, that one does not know where to begin to speak of it.

"The two main divisions of mathematics, analysis and geometry, correspond with some exactness to the two great mysteries of the Christian faith, the Trinity and the Incarnation."

CHAPTER - - - - - - - - - - - - - - 27

Return of the Master

THE first four decades of the twentieth century may be remembered as the beginning of a revolution in the world of thought no less profound than the accompanying political and social upheaval. Old ideas to which the mind had clung for centuries as necessities of rational thought—space, time, number, reason itself—were being modified beyond recognition for decades before the outbreak of the Second World War. The traditional universality of the "truths" of mathematics, also their assumed necessity, on careful scrutiny vanished, or became something more like other human knowledge. We have followed this change; we shall now note its counterpart in twentieth-century science.

By the middle 1930's it was evident that Pythagoras and Plato had succeeded in keeping their agreement to meet in 1920. Curiously enough Kant also made himself heard again, though he had not been invited to participate in the rejuvenation of science by either of the returned ancients. Although the sage of Königsberg had been repudiated long since by the mathematicians he found an unexpected welcome from some of the prominent modern Pythagoreans. His "apodictic truths" of mathematics, and his "synthetic judgments *a priori*," which the mathematicians had shown to be nonexistent in their domain, and which some of the modern logicians had declared existed nowhere else, suddenly materialized in novel disguises in physics.

Plato's reception was somewhat similar to Kant's. It was no news to the mathematical realists that the great philosopher was back. So far as they were aware he had never been away. But the common run of scientists were rather surprised to see him again. They mistakenly imagined that he left their territory for his own good and theirs, never to return, in the late sixteenth century.

With Pythagoras himself it was brilliantly different. He had hoped, but not seriously expected, to find mathematics as primitive as when he left it in the sixth century B.C. Nor did he. One bewildered glance over the shimmering expanse of twentieth-century mathematics convinced him that even his elastic numbers could not be stretched out far enough to cover everything he saw. Being a man of some practical sense in spite of all his numerology, the master decided to forget his numerical indiscretions and see what he could do in the region of his first great fame. To his inexpressible joy he discovered that several distinguished mathematical physicists and theoretical astronomers were going on where he had left off. He immediately joined them. The natural numbers 1, 2, 3, . . . and their ratios—the rational numbers—soon were again in full favor as the ultimate realities of the physical universe.

What he had always wanted to do himself, but had never been quite able to accomplish alone, was now easy with sympathetic modern minds as sieves to strain off all the impurities of his thought. By thinking hard enough and long enough about numbers and nothing else, the master squeezed through the most astounding scientific discoveries without making a single observation or performing one experiment. It was immaterial to him who might publish all these beautiful new epistemological necessities as his own. He, Pythagoras, knew that his, and his alone, was the ecstasy of creating them out of his pure reason, and his the certain knowledge that he had at last taken leave of his senses and was forever unbound from the Wheel of Birth.

But even in his moment of triumph the master sorrowed. His own disgraceful plucking of strings recurred to his memory, and he blushed. How could he ever have been so ignorant of the creative power of his own mind? Of course, he realized now when it was about twenty-four centuries too late to do his personal reputation any good, the law of musical intervals is an epistemological truism, a necessary consequence of the manner in which a rational mind must interpret the content of its sensory experience. Why had he not noticed this back in Croton? Was he really so stupid in that frightful incarnation when he tried to civilize Milo and his wife? Flushing with shame he suddenly realized why that hulking athlete and his commonplace little woman looked at him so queerly when they caught him picking at his monochord and measuring the length of the vibrating segment. At the time he had imagined them, in their rustic ignorance, thinking him daft. Now, all those disastrous centuries too late, he knew what they were too polite to say. Instead of mocking him for going to all that unnecessary labor to discover a necessary consequence of plain thinking, which they had known intuitively ever since they were weaned, they just stood in the doorway, saying nothing for fear of hurting a simple-minded guest's feelings. Could hospitality go farther? "No wonder," the master groaned, "the Wheel plunged me into all that mediaeval numerology. I deserved no less."

And so it happened once more, as so often in the development of science, mathematics, and philosophy, that methods and beliefs abandoned by leaders in one department of human knowledge were eagerly adopted by prominent men in another. Such returns to the past do not necessarily imply sterility or decadence. But the disinterested onlooker may wonder occasionally whether the latest adherents of ancient creeds, found untenable by others, know anything at all of the past of their newest enthusiasm. Perhaps it is well that sometimes they do not—there is nothing so effective as knowledge for paralyzing action.

Specifically, much of the scientific philosophy of the modern Pythagoreans seems to stem from the ancient confusion between pure mathematics, which is an abstract logical system empty of factual content, and applied mathematics, which is partly designed to accord with observable fact, and which to this extent is an empirical science. The tautologous vacuity of pure mathematics is transferred, subconsciously perhaps, to the mathematically formulated hypotheses and "laws" of the sciences; and with this factual emptiness, the illusion either of the eternal necessity or of the fictitious *a priori* character of mathematical truths, passes over into "all the laws of nature that are usually classed as fundamental."

The quoted phrase, the reader may recognize, is Eddington's, cited in our opening chapter. We now return to our starting point, and recall a few of the historical details which may have been partly responsible for the remarkable conclusion that these same fundamental laws of nature "can be foreseen wholly from epistemological considerations. They correspond to *a priori* knowledge, and are therefore *wholly subjective*." Kant, as we saw, held similar opinions regarding mathematical truths, especially those of geometry; and the theological logicians of the Middle Ages believed much the same about Aristotle's logic and rudimentary science. We saw also that the mathematicians of the nineteenth and twentieth centuries relinquished these opinions, for the sufficient reason that they are contradicted by modern knowledge. This, however, should not prejudice anyone against scientific Pythagoreanism. It is still under discussion by competent experts, and seems likely to remain there for many years to come. Let us admit once for all, before proceeding to the summary, that if the modern Pythagoreans are right, theirs is the least-anticipated and farthest-reaching scientific advance in twenty-five centuries.

Thales, Pythagoras, and their successors seem to be ultimately responsible for the current belief that it is possible merely by taking sufficient thought to discover all the fundamental laws of the physical sciences. Their elementary geometry originated in abstraction or idealization of sensory experience and the simplest observation of the world about them. Then, as we saw, they discovered that the truths of geometry are deducible from a few postulates or, as they came to be called, "common notions." The postulates appeared necessary, not merely sufficient, for a consistent account of the physical universe. Likewise the logic in the process of deduction seemed to be inevitable. It was but natural for metaphysical spectators to infer that the "laws of thought" and the "constitution of the mind" render all appeals to sensory experience not only superfluous but misleading.

Impressed by the triumphs of the geometers, the philosophers set up their own postulates—or sometimes concealed them—and proceeded to reason out what must be the constitution of the universe, the nature of the gods, and the relations of the human soul to both. The simple postulates again seemed necessary to their inventors or discoverers, and again the accompanying deductive reasoning appeared to be as rigid as fate—if not fate itself.

The philosophers were even more confident than the mathematicians of the necessary correctness of their conclusions; for obviously it was in general impossible to confront deduction with observation. In the closely reasoned abstract science generated by mathematics and philosophy it would have been feasible to check some of the deductions against fact. But on the whole the more influential leaders trusted their reasoning implicitly. This unquestioning reliance on the unaided pure reason as a necessary and sufficient implement of understanding and discovery passed from Greek science and philosophy into the orthodox scientific procedures of the European Middle Ages.

The great services which classical deductive reasoning rendered mediaeval theology lent that reasoning a spurious prestige in science. "All the saints and sages," discerning in the trivialities of elementary arithmetic the archetypal pattern of the universe, readily discovered all the secrets of nature in the inspired numerology of holy writ. The material world not being of much importance to the zealous men concerned primarily with the salvation of their own souls and those of others for the immaterial world beyond the grave, science was subordinated to theology in the works of the new masters. If observation and experiment contradicted reason, so much the worse for observation and experiment. Logic and theology united in confirming the Pythagorean verdict that number rules the universe.

Toward the close of this golden age of unquestioning faith, pedants and the more scholarly believers in the supremacy and all-sufficiency of the pure reason found an agreeable confirmation of their creed in the ancient—and therefore reputable—Ideas of the Platonists. Purged of its theological crudities, numerology no longer was suspect of absurdity repugnant to the learned mind. In its Platonic refinement the ancient magic of numbers was the very essence of natural science, as attested by respected men of science themselves. Then, almost in a day, with the advent of the modern scientific method in the late sixteenth century, even philosophic numerology ceased to fetter active men of science, and positive knowledge of the physical universe increased more rapidly than in any previous age.

A hint that the ancient numerology was merely in abeyance appeared toward the end of the eighteenth century in a solemn pronouncement by Laplace. This greatest disciple of Newton in mathematical astronomy was no empiricist, nor was he in any degree a critical mathematician. In fact, if it were not biographical blasphemy to speak the truth occasionally about the mighty dead, it might be candidly stated that outside his own special domain Laplace, both as a man and as a mathematician, was as naive as a child. This not unattractive intellectual innocence is but the more engaging in that the great mathematical astronomer conducted his private life with the shrewd cynicism of a French peasant. Occasionally he was so clever in trimming his beliefs to fit the politics of the moment that he found himself without a rag to cover his real convictions, if any. Something of this practical opportunism may have dictated his public utterances on the "sublime science" whose devoted and unselfish servant he professed to be. "Truth," he declared in an incautious confidence, "is my only master." So it is possible that he may have been seeking to impress the unmathematical public with the importance of his personal researches when he announced that his equations contained all the past history of "the world"—the solar system—and inexorably dictated its future. Since Newton's law of gravitation had been pronounced universal, it followed that the entire universe was a mechanically determined whole, governed solely by the eternal mathematics of the eighteenth century. Space was Euclidean; gravitation, everywhere and always, Newtonian; logic, in the main, Aristotelian; and mathematics was just about to enter the most creative period in its history. Laplace had spoken.

Not all of the outstanding mathematicians of the eighteenth century were as satisfied with themselves and with their own works as Laplace. The greatest of them all, Lagrange, tempered solid accomplishment with mild skepticism. Consequently he issued no resounding proclamations on the destiny of the universe. When teased to make a prophet of himself, Lagrange discouraged his annoyers with the simple declaration, "I don't know." But it was the more knowing Laplace who captured public opinion and incited others to out-pontificate him if they could. Several succeeded.

One in particular, Sir George Biddell Airy (1801-1892), merits immortality for his profound observation that the universe is a perpetual-motion calculating machine whose gears and ratchets are an infinite system of self-solving differential equations. Every atom in the universe exists solely because the equations of the universe endow it with being. In return for this nebulous gift of existence, the atom in its erratic motion undoes the equations certifying its existence. Airy's romantic mathematics of the cosmos was the nineteenth-century version of the ancient myth of mathematical permanence disguised in sensory experience as chaotic flux. Pythagoras was about to return.

It was physics that finally made Pythagoreanism acceptable to a certain type of modern scientific mind. To see how this came about we must review briefly some of the more spectacular predictions of nineteenth- and twentieth-century physics and astronomy. The predictions of mathematical physics and astronomy are of three kinds.

The first concerns a known effect and forecasts what its numerical measure will be under certain prescribed conditions. That is, the prediction is quantitative with reference to something already known qualitatively. Many of the experiments in any good laboratory manual of high-school physics are designed to conceal this type of prediction from the student. The beginner knows, for example, that light is reflected from a plane mirror, and he is required to verify the "law" that the angle of incidence is equal to the angle of reflection.

If he were acquainted with the mathematical theory of light he might dispense with the laboratory exercise, but would not if he had the elements of an orthodox physicist in him. This first type of prediction assigns a "measure"—a number—to a qualitative effect.

In the second and rarer kind of prediction, phenomena hitherto unobserved are foretold from the mathematical formulation of a theory. The prediction here is qualitative, and neither the theory nor the mathematics concerned may be sufficiently developed to forecast a measure of the new effect. The wave theory of light, for example, in its earlier stages might have predicted some of the observed facts about polarized light, but could not have supplied the subsequent quantitative account.

The third and rarest type of prediction combines the first two. Something qualitatively new is predicted and at the same time a quantitative estimate of the unobserved effect is assigned. When such predictions are verified in the laboratory they seem almost as miraculous as the successful efforts of the ancient prophets. In such instances the pure reason appears to the modern Pythagoreans to have revealed facts concerning the physical universe in whose discovery sensory experience had no part. This is the crux of the dispute. Is it really so that even the most mysterious of these predictions of the third kind was wholly independent of previous experience gained by the senses in the world of the senses? Opinion divides here.

The Pythagoreans contend that the predictions are independent of sensory experience: the mind, in making the prediction, is merely recovering from a hypothetical external universe what it put into that imagined universe in the delusion that it is observing something independent of itself. The other side suspects that without some observed data, however trivial, on which to base a mathematical (or epistemological) theory for any range of phenomena, the theory would necessarily be empty of all observable and factual content. To which the Pythagoreans reply that the experimental scientist, with his supposed empirical knowledge of the facts, can know no more of the "real world" than does a kitten chasing its tail. "What you put in, you get out," they reiterate, "no more, no less. So why all this fuss to discover with your senses what you might find in your own reasoning processes?"

Why, indeed, when we remember a few of the triumphs of the apparently unaided reason? Let us recall only five, three of which have already been described. These specimen predictions of the third kind have been selected from many because they offer an ascending sequence of unexpectedness. They also progress in seeming miraculousness with their historical order.

The first was Hamilton's prediction (1832), quickly followed by laboratory confirmation, of conical refraction. It had long been known that some crystals are doubly refracting: a ray of light in traversing the crystal is split and generally emerges as two. Hamilton predicted that under certain exceptional conditions, which he prescribed, the incident ray should emerge as a cone of rays—not merely as two, but as an infinity. His theory of systems of rays and the geometry of the wave-surface in a doubly-refracting medium led him to this conclusion, and enabled him to calculate the angle of the emergent cone.

Next (1846) the prediction of the planet Neptune, while perhaps not qualitatively as novel as Hamilton's discovery, was equally remarkable quantitatively. This success of Adams and Leverrier was discussed in earlier chapters.

The third prediction was of a different order. To appreciate its peculiar significance for the twentieth-century revolution in physics, we must remember that the second half of the nineteenth century was the great era of mechanical models of the physical universe. Light, for instance, was imagined as a transverse vibration of an elastic medium (the ether) filling all space, though no experiment succeeded in revealing any evidence that this hypothetical medium had more than a subjective reality. As one part after another of the model broke down under the impact of new knowledge, ingenious mechanics repaired the damage, and the machinery began creaking out its revised version of the universe. Each revision was a little more complicated and considerably more forced than the one before. Nobody thought of simply scrapping the unwieldy patchwork of abstruse mathematics and ad hoc hypotheses. Perhaps even Maxwell himself did not, when he quietly ignored it in the actual statement of his equations of the electromagnetic field (1861, 1864). The equations not only sufficed to describe a wide range of known phenomena; they also predicted the existence of wireless waves.

The new point of significance, not fully appreciated at the time, was the total absence of a scientific mythology to validate the equations. Here was a mathematical description that worked; why invent a mechanical model to explain it? A few academic attempts to fit Maxwell's equations into the mechanistic physics of its age failed. Adding nothing to either the descriptive or the predictive scope of the naked equations, all these scholarly efforts to confound simplicity, including one by Maxwell himself, were soon forgotten. Doubters who felt at ease with a theory only when it had been cramped into the mold of the Newtonian mechanistic physics, suspected Maxwell's prolific equations of some concealed indecency and shunned their company.

Two generations were to pass before Maxwell's revolutionary departure was recognized as the signal for a new era. Taking the scientific world by surprise, the modern quantum theory in 1925 deliberately abandoned all models—except the mathematical—of the physical universe. If a usable mathematical description of some part of physical science could be given, that sufficed. As Einstein was the first to insist, it is a waste of ingenuity to invent elaborate "unobservables" to account for the "observables" of nature. These alone are symbolized in the equations and are the sole objects of calculation and measurement. Viewed in retrospect, this disregard of unattainable "ultimate realities," so revolutionary when first proclaimed by its young prophets with all the fervor of youth crusading to save the world for sanity, seems like nothing more disturbing than a long-overdue tribute to common sense.

Maxwell's quiet revolt also foreshadowed the return of Pythagoras. Probably nobody in the 1860's could have foreseen that in turning its back on one scientific mythology, physics was facing another. The equations, which Maxwell had reached only after much thought and a minute study of Faraday's experimental researches in electricity and magnetism, are deducible from a simple induction from experience. The inductive conclusion here is negative. It was stated by Faraday and is familiar to every beginner in physics. After numerous failures to produce an electric field in the space within a hollow conductor by charging the outside of the conductor, it was stated as a postulate, abstracted from experience, that no charged hollow conductor can enclose an electric field. This postulate suffices for the deduction of Maxwell's equations. From it and some skill in mathematics it is possible to predict among many other phenomena the existence of wireless waves.

A sufficiently sophisticated intelligence might see the postulate as a necessity of the manner in which the mind must interpret its sensory evidence of "space"—in this instance the void within a closed surface—and the "real geometry" of the universe. But at present the ordinary scientific mind regards the postulate as an idealization from sensory experience, and imagines that it was come upon precisely as the first geometers arrived at their ideal straight lines: by abstraction from experience. Unlike the Platonic geometers however, the orthodox physicist does not perceive his electric postulate as the shadow of an Eternal Idea. Nor does he, like the Kantians and some of the modern Pythagoreans, ascribe it to the structure of the human mind.

The fourth and fifth examples of qualitative and quantitative scientific predictions have already been mentioned, but may be recalled here, as one precipitated the return of Pythagoras and the other induced him to linger indefinitely.

The relativistic theory of the gravitational field (1915-1916) predicted (among other things) that the lines in the spectrum of light emitted from a massive star (the Sun, for example) should be shifted toward the red end of the spectrum by a calculated amount, while the corresponding lines in light produced in a physical laboratory should show no such shift. The prediction was sustained by observed fact.

Likewise for the prediction (1927) of ortho-hydrogen and para-hydrogen by the modern quantum theory. Though sufficiently complex, the quantum theory offers one of the simplest and most plausible arguments that the laws of nature can be foreseen wholly from epistemological considerations. For the dependence of the quantum theory on empirical evidence for its basic postulates is apparently so slight, and actually so elusive, that anyone who wishes to overlook it may do so without seriously disturbing his scientific conscience.

The facts in the matter resemble those for the electrical postulate of the charged hollow conductor. One of the postulates in the quantum theory asserts that it is impossible to measure exactly at one time both the momentum and the position of a moving particle. If one of the measurements is made more precise, the other necessarily becomes correspondingly uncertain or inaccurate. The two are curiously connected (another postulate) by a numerical formula which need not concern us here. This "uncertainty principle" is the outcome of numerous failures either to execute or to imagine an experiment in which both the momentum and the position could be measured exactly at the same instant.

The Kantian-Pythagorean numerologists include this inability to conceive the desired experiment in their postulates concerning the nature of the human mind. To them it is a necessity of the way in which the mind must interpret the content of its sensory experience. Others class it as an induction from empirical evidence on a par with all scientific inductions. They deny that the extremely abstract general principles of modern physics are uncontaminated by observation and experiment, but admit—as anyone must—that the deductions by mathematical reasoning from these same rarefied principles are numerous, more frequently than not in accord with observation, and sometimes startlingly unexpected. They also find it hard to believe that it is possible to get something out of nothing, by mathematics or any other human means. To which the Pythagoreans agree, adding that what comes out is exactly that which was put in—by the mind.

Without mathematical technicalities it is impossible to show the scope and the power of such simple postulates of modern physics as this: If a system of bodies (say a cluster of stars) is moving with constant speed in a certain direction, it is impossible to detect the motion by observations performed wholly within the system (and therefore without reference to external bodies). This is one of the postulates of special relativity. A familiar illustration is the perfect railway car moving with unvarying speed along a straight track. If all the blinds are down the passengers cannot judge whether the car is at rest or in motion. But if the brakes are suddenly applied, or if the track passes into a curve, the decision is immediate.

Turning this relativistic postulate over and over in our minds we may easily persuade ourselves that it is a truism, perhaps a necessary logical consequence of the mere meanings of the key words "direction" and "constant speed." If we believe this we see that the postulate is more or less a matter of grammar and syntax or, if preferred, of semantics. What does it signify? It really says nothing about the world of sensory experience, except possibly in the assertion "it is impossible to detect." The last implies a detector—an observer —and he is supposed to be trying to do something. Ignoring him—he would have to perform "observations" forever to establish his "impossible"—we take the next obvious step and identify correct syntax with right thinking. It is then plain—perhaps deceptively so—that the postulate is a necessity of thought; the structure of our minds is such that we can conceive no alternative. We are therefore justified in believing that we have discovered one of the "epistemological principles" of physics.

Continuing in the same manner with all the recognized fundamental laws of the physical sciences, we find that several more of them might as well be posited—in the interests of economy of thought—as necessities of any consistent reasoning about the "external world." But if we happen to remember a little of the history of physics since Galileo and Newton, we may recall that all of these extremely powerful laws were attained only after decades of painstaking observation and laborious experiment. With all that drudgery happily behind us, we recognize now, say the Pythagoreans, that it was unnecessary. Provided our predecessors had introspected intensely enough they might have foregone all observation and all experiment. The truth (not an epistemological truism) that several of the Greek scientists and philosophers, and no few of the mediaeval logicians, did precisely this, and found but little that checked with sensory experience, is beside the essential point. Perhaps the moderns will be more successful, if only after the fact.

Even if the epistemological method in science should fail to find anything new, it may at least show that some of the old is more obvious than had been supposed. Any pruning away of superfluous hypotheses may be counted a gain. But it seems too much to expect that more than a half dozen of the leading scientists in any thousand years will master the introspective technique sufficiently to make new scientific discoveries. After all the principle—not the accompanying mathematical machinery—of general relativity might have been stated by Pythagoras himself. Yet Plato overlooked it, as did Aristotle, Newton, Maxwell, and hundreds of others who could have seen it.

The ultimate objective of the modern Pythagoreans is substantially the same as that of their ancient forerunners. They seek to discover a system of purely mathematical statements summing up everything knowable about the physical universe, and capable of predicting all physical events. Here "physical" is intended to exclude the living. The smaller the number of statements required the better; one is the ideal. All the "external world" will then be forever reduced to one grand mathematical formula. This combines the dream of Pythagoras and the ambition of Laplace in a unity beyond which there is nothing further to be discovered or imagined. But there is a difference, which Kant would have appreciated: the all-inclusive formula is to be sought and found in the structure of the mind. All the laws of the inanimate world will then be evident intuitively without appeal to the senses. Plato will not have lived for nothing.

A foretaste of what the outcome may be like was offered by Eddington in 1936, in his impressively suggestive *Relativity Theory of Protons and Electrons*. As the argument (329 large pages) is technical we can present only a few of the conclusions, referring those interested in forming an independent opinion to the work itself. There have been significant modifications of the theory since 1936, but none to destroy the characteristic features. New discoveries may be fitted in where necessary by a skilful addition of further assumptions as needed.

Eddington observes that there are certain recognized "constants of nature," seven of which are generally considered fundamental to physics and cosmology. Three, the mass of a proton, the mass of an electron, and the charge of an electron, are contributed by atomic physics; one, "Planck's constant," by the quantum theory; and three, the velocity of light, the "gravitational constant," and the "cosmical constant," by relativistic physics and cosmology. The mathematical expressions of these seven contain letters denoting arbitrary units of "length," "time," and "mass." By elementary algebra these arbitrary three may be eliminated. The seven constants thus generate an elemental and just "four" of pure numbers reminiscent of both Pythagoras and Empedocles. One of the four is a large number, N, which is asserted to be "the number of particles in the universe." Another, very famous, is the prime number 137, the basic "fine-structure constant" of spectroscopy. We shall return to 137 in a moment. Another is the ratio of the mass of a proton to that of an electron; this is a rational number. The proton and electron are two of the elementary particles out of which atoms are believed to be constructed. The remaining pure number furnished by the fundamental constants of nature is equally interesting but more technical.

The huge number N of particles in the universe, of course, has not yet been checked observationally. The other three pure numbers are reasonably small and all are well known. Thus an observational check for three of the four constants is feasible. The check is good, indeed better than good.

Eddington observes that "all four constants are obtained by purely theoretical calculation." He then remarks that the number (four) of dimensions of space-time—the framework of the physical universe, according to relativity—may be considered as a fifth fundamental constant of nature. "Even this number," he states, "is found to be determined unambiguously by the epistemological principle that we can only observe relations between two entities"—a principle which almost anyone might admit as a necessity of rational thought or meaningful language.

After noting the remarkable agreement between his own epistemological conclusions and the results previously known from observation and experiment, Eddington remarks that "It would have been disconcerting if it had turned out otherwise; but the theory does not rest on these observational tests." Further, if the theory is right, "it should be possible to judge whether the mathematical treatment and solutions are correct, without turning up the answer in the book of nature. My task is to show that our theoretical resources are sufficient and our methods powerful enough to calculate the constants exactly—so that the observational test will be the same kind of perfunctory verification that we apply sometimes to theorems in geometry."

Of all details of the epistemological theory the career of "137" is perhaps the most interesting. The fine-structure constant had been the subject of many experimental determinations (direct or indirect) before Eddington undertook to derive it from epistemological considerations. His calculations yielded 137 as the numerical value of this constant, slightly but significantly in disagreement with estimates obtained experimentally. The discrepancy between theory and observation was small enough to suggest more than coincidental agreement. Several competent experimenters repeated their work with meticulous care, or devised and applied new methods to test 137. Until the theory predicted that the constant must be an integer, and stated 137 as the integer, it had not been suspected that the constant might be a whole number. Pythagoras of course could have told the experimenters that they were bound to find an integer when they learned how to measure accurately. They did. By 1942 it was generally conceded that 137 is right.

Whether the epistemological theory stands in any form, modified or unmodified, or whether it falls, 137 will remain to its credit. The theory discharged its scientific obligation when it instigated new experimental work of great scientific value by any recognized standard. That the prediction was verified may turn out to have been only a lucky coincidence. But should this happen, it will not detract from the positive achievement. Nor will it be the first time in the history of science that the error of one man has been worth more than the truth of another.

When the new Pythagoreanism first appeared in 1920 it was ignored by all but a few physicists as a harmless mysticism of no significance for science. By 1937 it had acquired so considerable a following among men already distinguished for their positive scientific achievements that it could no longer be shrugged off. It was time to see exactly what "epistemological science" meant for the tradition of Galileo and Newton. Representatives of both the old and the new agreed to debate the matter so that the scientific public might know what each side stood for, and be able to form its own opinion. All the participants were recognized scientists who could speak with authority. As our parting tribute to the master we offer a few of their more interesting opinions.

The debate was opened by the theoretical astronomer, E. A. Milne, known for his "Cosmological Principle" which he proposed as a substitute for Einstein's relativity. According to Milne, "It is in fact possible to *derive* the laws of dynamics rationally . . . , without recourse to experience." These laws, we recall, are the foundation of physics according to Galileo and Newton, who obtained them by induction from experience.

The distinguished astrophysicist and philosopher of physical science, H. Dingle, championed the other side. "To the Aristotelian [a slip, for Platonic Pythagorean], the human mind had supersensory knowledge of the principles which Nature obeyed, or alternatively reason could, apart from sense, dictate the course of experience; to Galileo, Nature was independent, and the mind could watch and try to describe in general terms her processes, or alternatively reason could seek to correlate sense observations into a logical system." In contrast, the new Pythagoreanism exalts "cosmolatry—the idolatory of which 'the Universe' is god [which] transcends observation and cannot be derived from observation alone; it dominates rather than represents experience. This cosmolatry, as might be expected, came by metaphysics out of mathematics. . . . Thus we find among the general public a vague belief that physics is the study of the Universe, and in the scientific world a wholesale publication of spineless rhetoric the irrationality of which is obscured by a smoke-screen of mathematical symbols."

The language of the debaters was presently to become even

plainer and more robust. From time to time a milder remark, such as Pythagoras himself might have interjected, restored the dispute to the impersonal level customary in modern scientific controversies. We quote one such by P. A. M. Dirac, co-founder of the newer quantum theory, as a sample of what may be epistemologically attainable. The number written 10^{39} is 1 followed by 39 zeros. "We may take it as a general principle," Dirac asserted, "that all large numbers of the order 10^{39}, 2×10^{39}, 3×10^{39}, . . . , turning up in general physical theory are, apart from simple numerical coefficients, just equal to t, $t \times t$, . . . , where t is the present epoch expressed in atomic units. The simple numerical coefficients occurring here should be determinable theoretically when we have a comprehensive theory of cosmology and atomicity. In this way we avoid the need of a theory to determine numbers of the order of 10^{39}."

It may have been this mysterious hint of a new theory to supersede both physics and numerology that provoked one of the sharpest retorts in all the spirited debate. Refusing to be decoyed into a nebulous future by the bait of an inexistent but "comprehensive theory of cosmology and atomicity," Dingle recalled the debaters to the main issue. "But the question presented to us now," he reminded his opponents, "is whether the *foundation* of science shall be observation or invention. Newton did not lack imagination, but he chose to examine pebbles rather than follow the Gadarene swine, even when the ocean before him was truth. Milne and Dirac, on the other hand, plunge headlong into an ocean of 'principles' of their own making, and either ignore the pebbles or regard them as encumbrances. Instead of the induction of principles from phenomena, we are given a pseudo-science of invertebrate cosmythology, and invited to commit suicide to avoid the need of dying. It is the noblest minds that are o'erthrown, the expectancy and rose of the State which was lately so fair and in which there is now something so rotten that the very council of the elect [the Royal Society of London] can violate its charter and think it is doing science a service."

In case the allusions in this somewhat bellicose indictment of the Pythagoreans may have slipped anyone's mind we recall them. The ocean and the pebbles to which Dingle refers are those in Newton's estimate of himself near the close of his long life: "I do not know what I may appear to the world; but to myself I seem to have been only like a boy playing on the seashore, and diverting myself in now and then finding a smoother pebble or prettier shell than ordinary, whilst the great ocean of truth lay all undiscovered before me." The Gadarene swine, according to St. Mark, "ran violently down a steep place into the sea, and were choked in the sea," after "all the devils" that Jesus had cast out of a violent madman were given permission to enter "into the swine." But it is the Shakespearean thrust that really gets home. It would have been even more effective than it is if Dingle had included another line. Epistemological physics is the mad Hamlet who has just told Ophelia—presumably experimental science—"We will have no more marriages. . . . To a nunnery, go." To which the distracted Ophelia replies,

> "O, what a noble mind is here o'er thrown! . . .
> The expectancy and rose of the fair state,
> The glass of fashion and the mould of form,
> The observed of all observers, quite, quite, down!"

Or is it Milne and/or Eddington who, in repudiating "the observed of all observers," is Hamlet? The "something so rotten" of course has made its way into physics from Hamlet's "State of Denmark," the home of Niels Bohr, one of the more audacious innovators in twentieth-century theoretical physics. Though it may not be entirely clear who is who or what is what in Dingle's denunciation, any connoisseur must admire it as the little masterpiece of artistic cursing that it is. Not even that "great text in Galatians" with its "twenty-nine distinct damnations" compresses so much indignation into so narrow a compass. That a man of science can be so moved by a purely scientific matter is a good omen for the future of science. Physics will not die of excessive politeness so long as its admirers become excited about it, and express themselves as forcibly as some of them do over their partners' mistakes at bridge.

Ignoring the porcine personalities Milne soberly restated his position. In speaking of his own substitute for relativity he asserted that "It is an astonishing thing that the eliminaton of other empirical appeals, including all appeals to quantitative laws of physics, can be carried out as far as it can, however imperfect the present state of the theory. No one has been more astonished than [myself]. It is not an *a priori* belief to be scoffed at; it is a fact of experience to be reckoned with, that when we do thus eliminate such empirical appeals, regularities emerge (as logical consequences of [my] hypothesis) which play the part of the very laws of Nature which are *observed* to hold good. These regularities have the logical status of *theorems,* and the resulting logical structure has the status (or would have the status if it were perfect) of an abstract geometry based on axioms."

An attentive listener might have heard a subdued flutter of applause from at least two of the auditors, neither of whom had been officially asked to judge the debate, but both of whom volunteered their services. "I always told them so," Plato whispered, only to be met by the identical words from Kant. Out of respect to their common disciple they stopped whispering as Milne proceeded to analyze the problem of "the origin of the laws of Nature."

"Empirical physics," he declared, "cannot attack this problem." The problem, it appears, is "the belief that the universe is rational." It is therefore a modern echo of the dream of Pythagoras. Milne then elucidated his understanding of a real solution of the problem. "By this I mean that given the mere statement of *What is,* the laws obeyed can be deduced by a process of inference. . . . We can only test this belief by an act of renunciation, by exploring the possibility of deducing from some assumed description of just *What is* the nature of the laws which *What is* obeys, avoiding as far as possible all appeal to empirically ascertained laws. Laws of Nature would then be no more arbitrary than geometrical theorems. God's creation would be subject to laws not at God's disposal. The laws would be consequences of the world shape." Surely we heard some of this before from the Aristotelian logicians of the Middle Ages?

As might have been anticipated, Dingle's Gadarenes refused to be "choked in the sea" without a protest. Indeed several of them struck out boldly and safely reached dry land. After gracefully acknowledging "Dingle's entertaining article," Eddington "toned down its rhetoric a little" before attempting to demolish it. "Galilean" in his reply refers, as always in physics, to Galileo, not to anything in the source from which Dingle drew his uncomplimentary comparison. "My view," Eddington explained, "represents a definite contrast to the 'Galilean' view; and I feel a great satisfaction at having shocked a die-hard [Dingle] of the latter school. . . . After a rather extensive series of researches, I have found that a great part of the current scheme of physics is deducible by *a priori* argument and therefore does not constitute knowledge of an objective universe."

The murmur of approbation which interrupted this tribute to the *a priori* came from the direction of Kant. It passed unacknowledged as Eddington proceeded to his "N"—the impressive 2.136×2^{256} which in 1937 he deduced from his epistemological principles as the total number of particles in the universe. "When the quantum physicist states the number of particles in a system, whether few or many, he gives the number reckoned by quantum arithmetic. The natural constant N is a number in quantum arithmetic; it could have no other meaning, for Pythagorean arithmetic is a non-starter in the competition. . . . We find that in the corresponding [quantum] arithmetic the integers run only from 1 to 2.136×2^{256}. We can thus obtain the number of 'all the particles that there are' from our *a priori* knowledge of the arithmetic that is used for enumerating them. From the philosophical point of view we have debunked N."

To which Pythagoras might have replied that although his arithmetic—numerology—"is a non-starter in the competition," yet such is its perennial persistence that it easily wins the race over all competitors, starters and non-starters alike, as its distinguished repudiator had just demonstrated.

The assertive tone of the leading Pythagoreans did not pass unnoticed even by their sympathizers, and some tried to moderate it slightly. Thus Eddington's able coadjutor, the relativist W. H. McCrea, possibly feeling that too much had been claimed in the stress of debate, asked, "Is it then possible that we may dispense with all other hypotheses, that is, that all other hypotheses will be revealed by conventions of thought or the expression of thought? Eddington's theory . . . may be regarded as, in fact, an effort to do this. I fear, however, that I may have intruded into a region where angels fear to tread."

Newcomers, less timid, rushed into the debate, and the veterans also joined issue once more. Of those who had not yet participated, the Marxist biologist J. B. S. Haldane offered one of the most interesting contributions to the discussion—possibly because he could see Pythagoreanism from a vantage point inaccessible to the physicists. Skilled in the mathematics of genetics, and knowing the limitations of mathematical reasoning in the life sciences, Haldane could be more objective about scientific uses of mathematics than some who had experienced little else. The biologist dismissed epistemological physics and astronomy with the remark that Milne's "hypothesis would have appeared fantastic to Aristotle, Ptolemy, or St. Thomas."

Haldane was followed by H. Jeffreys, well known for his work on scientific inference, who offered his temperate diagnosis of modern Pythagoreanism in general. "I think," he hazarded, "the source of the trouble is the belief that there is some special virtue in mathematics. Instead of being regarded as what it is, a tool for dealing with arguments too complicated to be presented without it, it has become emotionalized to such an extent that many people think that nothing but mathematics has any meaning; whereas the opinion of some of the best pure mathematicians is that the characteristic feature of mathematics is that it itself has no meaning. . . . Its function is to connect postulates with observations." But, as we have seen, others of "the best pure mathematicians" still believe that "mathematical reality lies outside us."

Jeffreys' diagnosis was amplified by L. N. G. Filon, a mathematician and physicist of the older tradition: "The real trouble seems to be that, instead of starting from the facts of observation and gradually building up by induction particular laws which may or may not eventually be linked up, some men of science appear to think that they can solve the whole problem of Nature by some all-inclusive mathematical intuition. What they are really doing is not to explain Nature, but to explore the possibilities of the human mind. . . ." Then, a salutary touch of caustic: "I seem to remember a phrase to which a good deal of lip-service used to be paid, about hypotheses 'which were crushed in the solitude of the study.' Judging by much of the scientific literature which gets published nowadays, something seems to have gone wrong with the crushing machines."

It remained for a professional astronomer, R. A. Sampson, to remind the Pythagoreans that logic—theirs or that of anyone else—might have but little relevance for the factual world. "For where logic works," Sampson observed, "it offers to tell us what will happen in another time and place, of which, by hypothesis, we have no experience. Of course a large part of logic is explanatory [analytic, in Kantian], a mere unfolding of what is implied in the statements. Take, for example, mathematics. The statements found in Euclid are contained in the definitions, postulates and axioms. They are mere statements. None of them can be proved or disproved, and their interest is derived from the way in which they accord with the external world. All the rest is a process of unfolding; and so for other cases—all follow the same model. . . . It is the greatest error that mathematics makes that it slurs the distinction between the past and the future—the quantities that it handles are timeless."

Philosophers who ventured into the debate were somewhat rudely received by the distinguished mathematical physicist C. G. Darwin—almost as discourteously in fact as the classical metaphysicians had been welcomed out by the modern symbolic logicians. Darwin was even blunter than Carnap. "The fact remains," he pointed out, "that it is the science and not the philosophy that matters, and that most men of science do not find it worth their while to read the works of metaphysicians. Is it not the salient fact about the philosophy of science that no professional philosopher can write a book that a man of science wants to read? Ought there not to be a meta-metaphysics which would bring a message of comfort telling us not to bother about our philosophy—a command most of us already obey—because we can get on with our business without it? A book on this subject would show what a lot of things there are that do not matter much, and it would have the advantage that *a fortiori* no one at all need read it."

Well, perhaps. In any event the philosophers no doubt will give as good an account of themselves in the future as they did in the past.

The debate ended in a draw. Summing up, Dingle stated that "The criterion for distinguishing sense from nonsense has to a large extent been lost: our minds are ready to tolerate any statement, no matter how ridiculous it obviously is, if only it comes from a man of repute and is accompanied by an array of mathematical symbols in Clarendon type. . . . If this state of mind exists among men of science, what will be the state of mind of a public taught to measure the value of an idea in terms of its incomprehensibility? . . ." With the prescience of a twentieth-century Cassandra the disillusioned astrophysicist then nearly foretold the outcome of it all: "The times are not so auspicious that we can rest comfortably in a mental atmosphere in which the ideas fittest to survive are not those which stand in the most rational relation to experience, but those which can don the most impressive garb of pseudo-profundity. There is evidence enough on the Continent [of Europe] of the effect of the doctrines derived 'rationally without recourse to experience.' To purify the air seems to me an urgent necessity."

The foregoing dates from June, 1937. It may be only a coincidence, but the purifying wind, which in 1937 seemed so necessary to clear unreason out of the atmosphere, started to blow full blast from the Continent in September, 1939. It cannot be proved, of course, that a repudiation of experience, and a consequent return to doctrines derived without recourse to it, were of any significance in the world of practical affairs and brute experience. Those who insist that all science beyond that of the machine shop and the arsenal is of purely academic interest, of no importance for the mass of mankind, may be right. But if the past is significant this seems unlikely.

The wind rose, but the air was not immediately cleansed. Hearing the approaching storm, Pythagoras remembered Croton and Sybaris, and was downcast. The dream he had gone through purgatory to preserve seemed about to be swept back into the past forever.

But not all was being obliterated. The mathematics he had nourished as a seedling was now a furiously growing tree overshadowing the battlefields and munition factories of the world. Without mathematics, no science; without science, no armaments; without armaments, a more degrading slavery, perhaps, than any that the master's own feeble science, at last mature, might have abolished. Must he be bound to the Wheel again for one last turn? "Croton and Sybaris, Sybaris and Croton, farewell and hail!

> 'The world is weary of the past,
> Oh, might it die or rest at last!' "

A CATALOG OF SELECTED

DOVER BOOKS

IN ALL FIELDS OF INTEREST

A CATALOG OF SELECTED DOVER BOOKS IN ALL FIELDS OF INTEREST

CATALOG OF DOVER BOOKS